CAMBRIDGE LIBRARY COLLECTION

Books of enduring scholarly value

Zoology

Until the nineteenth century, the investigation of natural phenomena, plants and animals was considered either the preserve of elite scholars or a pastime for the leisured upper classes. As increasing academic rigour and systematisation was brought to the study of 'natural history', its subdisciplines were adopted into university curricula, and learned societies (such as the London Zoological Society, founded in 1826) were established to support research in these areas. These developments are reflected in the books reissued in this series, which describe the anatomy and characteristics of animals ranging from invertebrates to polar bears, fish to birds, in habitats from Arctic North America to the tropical forests of Malaysia. By the middle of the nineteenth century, this work and developments in research on fossils had resulted in the formulation of the theory of evolution.

The Great Auk, or Garefowl

The great auk (*Pinguinus impennis*, formerly *Alca impennis*), a flightless bird of the north Atlantic, became extinct in the mid-1850s because of over-hunting – apart from being used as a food source and as fish-bait, its down was used for feather beds, and efforts in the early nineteenth century to reduce the slaughter were not effective. The last breeding pair was killed in 1844. This 1885 work by Scottish naturalist and scientist Symington Grieve (1850–1932) collects together 'a considerable amount of literature bearing upon the 'History, Archaeology, and Remains of this extinct bird'. The material includes articles on the historic distribution of the great auk, its known habits, its various names, and information on all the surviving specimens, whether stuffed, skeletal, bones, or eggs. The book is illustrated with drawings and lithographs of auk remains, and an appendix supplies historical and contemporary documents on the auk from all over Europe.

Cambridge University Press has long been a pioneer in the reissuing of out-of-print titles from its own backlist, producing digital reprints of books that are still sought after by scholars and students but could not be reprinted economically using traditional technology. The Cambridge Library Collection extends this activity to a wider range of books which are still of importance to researchers and professionals, either for the source material they contain, or as landmarks in the history of their academic discipline.

Drawing from the world-renowned collections in the Cambridge University Library and other partner libraries, and guided by the advice of experts in each subject area, Cambridge University Press is using state-of-the-art scanning machines in its own Printing House to capture the content of each book selected for inclusion. The files are processed to give a consistently clear, crisp image, and the books finished to the high quality standard for which the Press is recognised around the world. The latest print-on-demand technology ensures that the books will remain available indefinitely, and that orders for single or multiple copies can quickly be supplied.

The Cambridge Library Collection brings back to life books of enduring scholarly value (including out-of-copyright works originally issued by other publishers) across a wide range of disciplines in the humanities and social sciences and in science and technology.

The Great Auk,
or Garefowl

Its History, Archaeology, and Remains

Symington Grieve

CAMBRIDGE
UNIVERSITY PRESS

CAMBRIDGE
UNIVERSITY PRESS

University Printing House, Cambridge, CB2 8BS, United Kingdom

Cambridge University Press is part of the University of Cambridge.

It furthers the University's mission by disseminating knowledge in the pursuit of
education, learning and research at the highest international levels of excellence.

www.cambridge.org
Information on this title: www.cambridge.org/9781108081474

© in this compilation Cambridge University Press 2015

This edition first published 1885
This digitally printed version 2015

ISBN 978-1-108-08147-4 Paperback

This book reproduces the text of the original edition. The content and language reflect
the beliefs, practices and terminology of their time, and have not been updated.

Cambridge University Press wishes to make clear that the book, unless originally published
by Cambridge, is not being republished by, in association or collaboration with,
or with the endorsement or approval of, the original publisher or its successors in title.

The original edition of this book contains a number of colour plates,
which have been reproduced in black and white. Colour versions of these
images can be found online at www.cambridge.org/9781108081474

ENGRAVED FIGURE OF A STUFFED SPECIMEN OF THE GREAT AUK
In the Central Park Museum, New York.

THE GREAT AUK, OR GAREFOWL

(*Alca impennis*, Linn.)

Its History, Archæology, and Remains

BY

SYMINGTON GRIEVE

EDINBURGH

LONDON

THOMAS C. JACK, 45 LUDGATE HILL

EDINBURGH: GRANGE PUBLISHING WORKS

1885

𝔅𝔞𝔩𝔩𝔞𝔫𝔱𝔶𝔫𝔢 𝔓𝔯𝔢𝔰𝔰
BALLANTYNE, HANSON AND CO.
EDINBURGH AND LONDON

This Work

IS

DEDICATED BY PERMISSION

TO

THE LEARNED AND VENERABLE SCIENTIST,

J. JAPETUS S. STEENSTRUP,

DOCTOR OF PHILOSOPHY AND MEDICINE,

PROFESSOR OF ZOOLOGY IN THE ROYAL UNIVERSITY, COPENHAGEN,

WHO,

AS THE FIRST TO WRITE A MONOGRAPH ON *ALCA IMPENNIS*, Linn.

MAY BE DESIGNATED

THE FATHER OF GAREFOWL HISTORY.

PREFACE.

IN submitting these pages to the public, the Author has fears that they will not bear severe criticism; but he must plead as some excuse that they have been compiled during the relaxation of evenings that have followed the toils of active business life. If chance circumstances had not led him to devote some study to the subject of *Alca impennis*, Linn., and in course of time brought within his reach a considerable amount of literature bearing upon the History, Archæology, and Remains of this extinct bird, it is most improbable he would ever have undertaken this Work. As his studies progressed he was led to suppose that it might be of some use to Ornithologists, if not also to a number of general readers, if he were to publish the information collected, as no detailed work on the subject existed, and the scattered notices regarding *Alca impennis* principally to be found in the Publications of the learned Societies are difficult of access.

The Author is deeply sensible of the obligations he is under to home and foreign *savants* for the information they have so willingly given, as it has enabled him to make his Work much more complete than he at one time supposed was possible, and also to give all the latest information. To Professor J. Steenstrup, Copenhagen, he is indebted for the valuable remarks which appear throughout the Work, and for permission to give translations of portions of his writings on *Alca impennis*, Linn.; also for his kindness in going over all the proofs. To Professor W. Blasius, Brunswick, he would tender his best thanks for allowing an epitomised translation of his recent Publication on the Remains of *Alca impennis*, Linn., to be prepared, and also for going over the proofs of that translation, and giving additional and more recent information. To Robert Champley, Esq., Scarborough, he is under great obligations for favouring him with the use of interesting correspondence, and also for giving him valuable hints and information, besides going over all the proof-sheets.

To Professor A. Newton, Cambridge; Dr. R. H. Traquair, Alexander Galletly, Esq., and John Gibson, Esq., all of the Museum of Science and Art, Edinburgh; Dr. J. Murie, of the Linnæan Society, London; John Hancock, Esq., Newcastle-on-Tyne, and many others, he is indebted for assistance and information, and desires to express to these friends and correspondents his sincere thanks. There is one gentleman to whom the Author is under greater obligations than any other, and he is the friend who has made the translations, revised the manuscript, and then the proofs, but at his own request he will be nameless.

With regard to the Illustrations, the Author desires to express his thanks to the President and Council of the Linnæan Society, London, for kindly granting him the use of the electrotype from which the picture of Caisteal-nan-Gillean, Oronsay, has been printed, and also for the use of the stone from which the plate of Great Auk bones found in the same shell-mound has been lithographed. He is also under obligations to the President and Council of the Scottish Society of Antiquaries for their kindness in giving the electrotypes from which have been printed the figures of the Great Auk bones found at Keiss, in Caithness-shire, and also the reduced reproduction of the figure of the Great Auk in the "Museum Wormianum, seu Historiæ Rerum Rariorum."

To the Authorities at the Museum of Science and Art, Edinburgh, he is indebted for the facilities afforded to Messrs. Banks & Co. for the execution of the drawings of the eggs of *Alca impennis*, Linn., from which the coloured plates given at page 108 have been prepared.

For the drawing of the only bone of the Great Auk yet found in England, the Author is under obligations to John Hancock, Esq., Newcastle-on-Tyne.

7 QUEENSBERRY TERRACE,
EDINBURGH, *July* 1885.

CONTENTS.

———◆———

APPENDICES.

ILLUSTRATIONS.

THE GREAT AUK, OR GAREFOWL.

CHAPTER I.

INTRODUCTION.

THE following pages have been written in the hope of interesting some in the story of an extinct bird. The whole history of the Great Auk is a sad one —the continued slaughters of the helpless victims culminating in the final destruction of the race on the skerry, named Eldey, off the coast of Iceland, excites to pity. The last of the Great Auks has lived and died. The race was blotted out before naturalists, when too late, discovered it was gone. Regrets are now useless—the living Garefowl is extinct.

Mankind, when the prize they value has passed from their grasp, wish it back again. But the Great Auk has gone for ever, and has left but few of its remains to recall its existence to the recollection of the future naturalists of the world.

Forty or fifty years ago, it was only among a small circle of ornithologists that the Great Auk was known and acknowledged. If in wider circles it had been heard about, it was looked upon as a myth. Now it is acknowledged by all, and has afforded perhaps more material for discussion than any other British bird.

Though much has been written and published in Britain and the Continent during the last thirty years, with the view of putting on record what is known regarding the Great Auk, all the information regarding it has, with the exception of the necessarily very meagre accounts given in popular works on natural history, appeared in the private publications of some of the learned societies. We therefore propose to go into greater detail in these pages, not with the impression that we have much to relate that is new to British ornithologists, but more with the desire to bring within the reach of all, materials that at present are difficult of access.

A

The first writer who wrote a Memoir of the Garefowl or Great Auk was the veteran Professor J. Steenstrup of Copenhagen. His paper was published in 1855, and we shall have repeated occasion to refer to it in the course of the following pages.[1] But previous to that time two workers were in the field, who visited Iceland, one of them more than once, with the object of ascertaining all that was possible regarding the Garefowl, especially the last scenes of its life. Those gentlemen, the late Mr. J. Wolley and Professor A. Newton of Cambridge, did good service to science by their labours ; and the latter has contributed, in several papers,[2] the results of their united work, besides putting on record much that has come to his knowledge from other sources. It is to be hoped that he may be spared to write a full history of " The Great Auk and its Remains," for which he has capabilities possessed perhaps by no other ornithologist at the present time.

Among other writers who have contributed papers on this bird are Mr. R. Champley[3] of Scarborough, and two well-known Scotch naturalists, namely, Mr. Robert Gray,[4] and the late Dr. John Alexander Smith of Edinburgh.[5] Of Continental authors there may be mentioned, Professor W. Preyer of Jena,[6] Mons. Victor Fatio of Geneva,[7] and Professor Wh. Blasius of Brunswick,[8] all of whose contributions are exceedingly valuable. Professor Owen has published an account of its osteology.[9]

We believe that the first note of warning that the Great Auk was likely to become extinct was sounded by a writer in a Danish journal during the year 1838. He says—" The Garefowl is likely to become extinct, like the Dront and other birds. There is a tradition that it has been seen in the Cattegat in earlier times. It has now, on the other hand, long disappeared from the coasts of Norway, Faröe, and Iceland."[10] This reference to Iceland, where the bird still lingered, reduced

---

[1] " Et Bidrag til Geirfuglens Naturhistorie," &c., in " Videnskabelige Meddelelser," fra den naturhistoriske Forening i Kjöbenhavn for Aaret 1855 (Copenhagen, 1856–1857), pp. 33–116. (With a plate and map.)

[2] *Ibis*, vol. iii., 1861, p. 374. Mr. J. Wolley's "Researches." " The Garefowl and its Historians," Natural History Review, 1865, p. 468, &c. &c.

[3] "Annals and Magazine of Natural History," 1864, vol. xiv. p. 235, &c.

[4] "Birds of the West of Scotland," 1871, p. 441–453. " Proceedings of Royal Society, Edinburgh," 1879–80, p. 668, &c. &c.

[5] " Proceedings Scottish Society of Antiquaries, Edinburgh," 1878–79, pp. 76–105 ; also 1879–80, pp. 436–444, &c. &c.

[6] " Ueber *Plautus impennis*," published at Heidelberg, 1862.

[7] " Bulletin de la Société Ornithologique Suisse," tome ii., parts 1 & 2, 1868.

[8] " Zur Geschichte der Ueberreste von *Alca impennis*, Linn." Naumburg, 1884. (See Appendix II.)

[9] "Transactions of the Zoological Society," London, vol. v. p. 317, 1865.

[10] " Naturhistorisk Tidskrift," 1838–39, p. 207. Writing 30th March 1885, Prof. J. Steenstrup informs us, " The writer was the editor of the ' Tidskrift,' but he here quotes from the ' Skandinavisk Fauna,' ii., 1835, p. 523, the work of the celebrated ornithologist, Professor S. Nilsson."

to a very small colony, attracted the attention of Professor J. Reinhardt, who wrote an article for the same journal during 1839 [11] on " Garefowl Appearances in Iceland." The learned Professor was thoroughly alive to the destruction of Garefowls that had occurred at their only hatching place during the years that immediately succeeded 1830, yet he does not appear to have thought that the death of the last of the Garefowls was so near, as only five short years were to elapse from the time he penned his communication until the fate of the Garefowl was sealed in the death of the last of the race.

In the following pages, which are principally devoted to the Archæology and History of the bird, we endeavour to place before the reader all the information which has come under our notice that is of most importance, though if an attempt were made to repeat all that has appeared regarding the existence of this bird, especially in the American region, we might fill several volumes. Whilst we have been thus careful to give a summary of all we have been able to ascertain regarding the existence of the Great Auk from the earliest times, the last scenes in its history during the present century are reserved for fuller details.

---

[11] " Naturhistorisk Tidskrift," 1838–39, p. 533.

CHAPTER II.

THE DISTRIBUTION OF THE GREAT AUK.—THE LIVING BIRD IN ITS AMERICAN HABITATS.

IN order to trace out the area in the northern hemisphere in which the Great Auk existed, it will be necessary to bear in mind that the only mode of doing so is to find out if possible the localities in which it bred, the recorded occurrences of its observation or capture, and lastly, the stations at which its remains have been recovered.

It is quite possible that our knowledge of what were the breeding places of the bird may be defective; but it seems the following are historically well attested, —namely, St. Kilda, Orkney, possibly Shetland, Faröe, the three Garefowl rocks off the coast of Iceland, Danells or Graahs islands situated in latitude 65° 20′ N., at one time called Gunnbjornsskjoerne;[1] then we have to go west to the east coast of North America, where, in the neighbourhood of Newfoundland, it was met with on Funk and many other islands;[2] also on some of the islands in the Bay of St. Lawrence, and at Cape Breton; while another station on the same coast at which it probably occurred was Cape Cod, the latter apparently being about the southern limit of the region in which the bird lived.[3]

Whatever may have been the numbers of the Garefowl in the eastern region during prehistoric times, the bird does not appear to have attracted the attention of the earliest writers. If it ever existed very numerously in this locality, it

---

[1] "Grönlands Historiske Mindesmærker," vol. i. pp. 123, 124. In addition to what is stated in the work we quote, we may mention that some centuries ago there were Norse settlements on the south-west coast of Greenland, but a sudden change in the climate of the country occurred, generally supposed to have been caused by an alteration of the current of that portion of the Gulf Stream which is believed at one time to have beat upon this shore. The sudden lowering of the temperature compelled the Norsemen to leave this part of Greenland, and the existence of those settlements had almost been forgotten until the discovery of the remains of their villages within recent times. It was possibly the same change in climate that caused the Great Auk to leave East Greenland, as there is no record of its having been destroyed at Gunnbjornsskjoerne, though the writer we refer to mentions its occurrence in large numbers about the year 1652. See " Ueber Plautus impennis," pp. 22, 23. See also Prof. Steenstrup's remarks in our Appendix IV.

[2] "Hakluyt's Collection of Voyages," London, 1600, pp. 133, 162, 173, 194, 195, 200, 202, 203, 205, 212. " Et Bidrag til Geirfuglens," " Videnskabelige Middelelser," 1855, No. 3–7, p. 95.

[3] " Videnskabelige Meddclelser," 1855, No. 3–7, p. 96.

would fall an easy prey to the primitive races of men, who, valuing it as food, doubtless killed all that came within their reach, until the few colonies of the bird that remained were confined to outlying islets, seldom wandering from their neighbourhood except when forced by circumstances, such as occurred through a volcanic subsidence off the coast of Iceland in 1830.[4] Its food being fish, which were more easily obtained on the banks or shallow water usually found near the land, seems to have generally confined this bird within soundings,[5] and this idea was held by sailors and fishermen, and is apparently based on accurate observation. We believe that the earliest notices of the Garefowl (or Penguin, as it was called in the American locality) are to be found in the works of writers, who, referring to the early voyages to the North American waters or the fisheries at the Banks of Newfoundland, mention the immense numbers of these birds.[6] Here also the Garefowl, probably long before the arrival of Europeans, had become confined to islets to which it could not be followed by the Red Indian in his frail canoe ; who would likely have no knowledge of the existence of such multitudes of birds ; though in later times, from the information gained by contact with white men, and his becoming possessed of boats, he annually visited the Bird islands for supplies.[7] In recent times, like the Great Auk, these aborigines have also become extinct. But if the Garefowl was safe from the red man in early times, it found a dreadful enemy in the white ; and the record of the war of extermination which he waged begins in 1497 or 1498, and went on until no Penguins were left to kill.[8] So valuable did these Garefowl prove as an article of food, that the ships which frequented the Banks for fishing were principally provisioned with them, as they fell an easy prey to the mariners, and were so stupid when on land that they allowed themselves to be driven on board the vessels on planks or sails spread out from the sides of the ships to the shore. Another plan which was resorted to, but probably in later times when the bird became less plentiful, was to drive them into compounds, where they were slaughtered with a short stick or club.

The following extract, taken from the notice in " Hakluyt's Voyages " (vol. iii. p. 130, ed. 1600, London), of the " Voyage of M. Hore and diuers other gentlemen to Newfoundland and Cape Breton in the yeere 1536," well describes a visit

---

[4] *Ibis*, vol. iii., 1861, p. 380. Mr. J. Wolley's " Researches."

[5] Pennant's " British Zoology," ed. 1812, vol. ii. p. 147.

[6] Edward Haie's " Report of Sir Humphrey Gilbert's Expedition to Newfoundland," 1583 A.D. " Hakluyt's," vol. iii. 1810, pp. 184–203, &c. &c.

[7] " A Journal of Transactions and Events during a Residence of nearly Sixteen Years on the Coast of Labrador," by George Cartwright, under the date, Tuesday, July 5th, 1785.

[8] " Hakluyt's Voyages," vol. iii. (Sebastian Cabot). London, 1600, p. 9.

to Penguin island and what they saw and did : " From the time of their setting out from Grauesend they were very long at sea, to witte, aboue two moneths, and neuer touched any land untill they came to part of the West Indies about Cape Briton, shaping their course thence northeastwards untill they came to the *Island of Penguines,* which is very full of rockes and stones, whereon they went and found it full of *great foules, white and grey, and big as geese, and they saw infinite numbers of their egges.   They draue a great number of the foules into their boates upon their sayles, and took many of their egges, the foules they flead, and their skinnes were very like hony combes full of holes ; being flead off, they dressed and eate them, and found them to be very good and nourishing meate.*"

This statement gives a pretty accurate idea of the numbers of the Garefowl and the ease with which they were captured ; and as the Newfoundland fisheries became developed, and vessels made regular trips to the Banks, they appear to have made very limited provision for the crews, depending upon the Garefowl to make up all deficiencies in their larders.   The following reference (taken from " Hakluyt," London, 1600, vol. iii. p. 133), illustrates how the practice was carried out.   It is part of a letter from M. Anthonie Parkhurst, gentleman, dated from Bristow, 13th November 1578, and addressed to M. Richard Hakluyt of the Middle Temple : " *There are sea Gulls, Murres, Duckes, wild Geese, and many other kind of birdes store, too long to write, especially at one island named Penguin, where wee may driue them on a planke into our ship as many as shall lade her.   These birds are also called Penguins and cannot flie ; there is more meate in one of these then in a goose ; the Frenchmen that fish neere the grand baie, doe bring small store of flesh with them, but victuall themselves always with these birdes.*"   There seems to have been no restriction put upon those men, and possibly many birds were needlessly killed and their bodies not even removed ; and it may bo to this cause that we owe the discovery during 1863 and 1864 of several mummy specimens, which were dug out from the frozen soil of Funk Island, two of which are now in England prepared as skeletons, one is in the British Museum.   Some of the bones in the mummy from which this skeleton was prepared were awanting, but the deficiency was fortunately made up through the kindness of Mr. J. Hancock, Newcastle-on-Tyne, who succeeded in extracting the bones similar to those that were missing from a skin in his possession.[9]   The other skeleton is now in Cambridge.   (See also pp. 28, 82.)

---

[9] *Ibis,* 1865, p. 116.

As the Garefowl appears to have laid only one egg each year, it may be easily understood that its reproduction would be very·slow, and that it could not long resist the war of extermination waged against it. Consequently we find that it gradually became fewer in numbers at all the American breeding-places, until finally, early in the present century, it altogether disappeared; and Professor A. Newton thinks if any are in existence, the only place where may possibly linger the last of the American Garefowls is the Virgin Rocks near the edge of, and midway on the north-west side of, the Great Bank off the coast of Newfoundland.[10]

It is stated that Colonel Drummond Hay, in passing over the tail of the Newfoundland Banks in December 1852, saw what he believed to be a Great Auk. This gentleman also sent Professor A. Newton a letter which he had received in 1854 from the late Mr. J. Macgregor of St. John's, Newfoundland, in which he states that in the preceding year, 1853, a dead one was picked up in Trinity Bay. But inquiries instituted by Professor Newton regarding this specimen did not result in any further information about it being obtained.[11]

Audubon mentions that Mr. Henry Havel, the brother of his engraver, while on a voyage from New York to England, hooked a Great Auk on the Bank of Newfoundland, in extremely boisterous weather; and also that when he (Audubon) was visiting the coast of Labrador, the fishermen stated that the Great Auk still bred upon a low rocky islet to the south-east of Newfoundland, where great numbers of the young were destroyed for bait; but as this information was received too late in the season, he had no opportunity of ascertaining its accuracy.[12]

More than thirty years have now elapsed since the last reported observation in the American locality, and as each succeeding year goes past without any notice of its existence, the hope must gradually die out among ornithologists that any of the birds have escaped. We may add, that from what is now known it is almost certain that all reported observations of the Great Auk since 1844 are mistakes.

---

[10] "The Garefowl and its Historians," by Professor A. Newton, Natural History Review, 1865, p. 486.

[11] Mr. J. Wolley's "Researches," by Professor A. Newton. *Ibis*, 1861, vol. iii. p. 397.

[12] "Ornithological Biography," 1838, p. 316.

CHAPTER III.

THE LIVING GAREFOWL IN ITS EUROPEAN HABITATS.

IN the European region the Garefowl in historic times is not known to have been ever as numerous as it was found by the early voyagers to American waters. But it certainly occurred in strong colonies at one or two stations, such as St. Kilda,[1] Iceland,[2] and probably the Faröe[3] and the Orkney Islands;[4] but from similar causes to those which operated elsewhere, it gradually was killed off, until in 1844,[5] or possibly 1845,[6] the last was heard of the living Garefowl.

ST. KILDA.

Of these birds the latest seen at St. Kilda was captured during the early summer of 1821 by two young men and two boys, who were in a boat on the east side of the island, and observed it sitting on a low ledge of the cliff. The two young men were landed at opposite points of the ledge, but about equidistant from the bird, which they gradually approached, whilst meantime the boys had rowed the boat close up to the rock under where it was sitting. At last, becoming frightened by the approach of the men, it leaped down towards the sea, but only to fall into the arms of one of the youths, who held it fast. Five years ago (1880) one of these boys, Donald M'Queen, was still living, aged 73.[7] From these men the bird was obtained by Mr. Maclellan, the tacksman of Glass or Scalpa, one of the Northern Hebrides. To this gentleman has been given the credit of having captured it, through some misapprehension on the part of the

---

[1] "A Voyage to St. Kilda," by M. Martin, Gent., London, 1753, p. 27.

[2] Mr. J. Wolley's "Researches in Iceland," *Ibis*, vol. iii., 1861, pp. 374–398.

[3] Olaus Wormius in his "Museum Wormianum, seu Historiæ Rerum Rariorum" (Copenhagen), Leyden, 1655, p. 301.

[4] MacGillivray, "British Birds," vol. iv. p. 361; and Appendix to the Supplement, Montagu's "Ornithological Dictionary," 1813.

[5] "Et Bidrag til Geirfuglens," by Prof. Steenstrup, "Videnskabelige Meddelelser," 1855, Nr. 3–7, p. 78.

[6] Thomson's "Birds of Ireland," vol. iii. p. 239.

[7] "Proceedings of Royal Society, Edinburgh," 1879–80, p. 669. Note "Letter from R. Scot Skirving, Esq., 17th June 1880, to Robert Gray, Esq."

Rev. John Fleming, D.D., minister of Flisk, afterwards Professor Fleming, of the New College, Edinburgh, who obtained it from him on the eve of his leaving Glass in the yacht of the Commissioners of the Northern Lighthouses, 18th August of that year.[8]

Dr. Fleming states, "The bird was emaciated, and had the appearance of being sickly, but in the course of a few days became sprightly, having been plentifully supplied with fresh fish, and permitted occasionally to sport in the water with a cord fastened to one of its legs to prevent escape. Even in this state of restraint it performed the motions of diving and swimming under water with a rapidity that set all pursuit from a boat at defiance. A few white feathers were at this time making their appearance on the sides of its neck and throat, which increased considerably during the following week, and left no doubt that, like its congeners, the blackness of the throat-feathers of summer is exchanged for white during the winter season."[9] The year in which this event took place has been supposed by some to have been 1822, as, owing to a misprint or mistake in the "History of British Animals," published by Professor Fleming in 1828, the latter date is given. But as the bird was obtained during a tour of inspection with the Northern Lighthouse Commissioners, an examination of their Journal has shown that the Rev. Dr. Fleming of Flisk was on board their yacht *Regent* in 1821, but not in 1822, when the Commissioners visited the island of Scalpa.[10] Unfortunately, the Garefowl escaped when the yacht was near the entrance to the Firth of Clyde,[11] as it was being allowed to take its usual bath in the sea with a cord attached to its leg; and there appears to be some evidence that this bird afterwards died, and its body cast ashore at Gourock.[12] The escape seems to have occurred after Dr. Fleming and his party had left the yacht, as they landed at the lighthouse at the Mull of Cantyre on 26th August, and proceeded by land to Campbeltown, whence they got the steamer for Glasgow. This bird had been given to Mr. Stevenson, the engineer of the Northern Lighthouse Board, and he gave it to Dr. Fleming, on the understanding it was to be presented to the Museum of Edinburgh University, and its unfortunate loss is perhaps irreparable.

---

[8] "Proceedings of the Society of Antiquaries of Scotland," vol. ii. n.s., p. 441.
[9] "Edinburgh Philosophical Journal," vol. x., 1824, p. 94.
[10] "Proceedings Society of Antiquaries of Scotland," vol. ii. n.s., p. 441.
[11] "Edinburgh Philosophical Journal," vol. x., 1824, p. 95.
[12] "Birds of the West of Scotland," R. Gray (1871), pp. 441-453.

ORKNEY AND SHETLAND.

The last notice of this interesting bird appearing in either Orkney or Shetland was in 1812, when two, a male and female, were killed at Papa Westra.[13] One account says that the female was seen sitting upon the rocks close to the sea, where it was knocked over by some boys or young men with stones, but was not then obtained. It was washed ashore some time afterwards.[14] Another account says it was shot. The male was chased by Mr. Bullock in a six-oared boat for many hours, but its speed was so great that the pursuit had to be abandoned. However, this bird was afterwards captured by some fishermen, who killed it and sent the body to that gentleman.[15] At his death it was sold, and purchased by Dr. Leach for £15, 5s. 6d., and placed in the British Museum,[16] where it is now the finest specimen they possess.

FARÖE ISLANDS.

At Farÿe the Garefowl had become exceedingly rare at the beginning of the present century,[17] and the last birds were probably killed during the next few years, but the exact date of the last capture cannot be ascertained. When Mr. Wolley visited Farÿe in 1849, he was told by an old man that he had seen one sitting upon some low rocks about fifty years before.[18] Professor J. Steenstrup relates, that during his visit to these islands he saw the head of a Great Auk which had been preserved. It is probable that either the last, or among the last, of these birds killed in this locality is referred to by Graba, who was at Farÿe in 1828. He found that most of the natives did not know it even by name; but some old people told him they thought they had formerly seen it at Westmannshavn, and one man said that he had killed an old Garefowl with a stick as it sat on its egg at this place.[19] From a remark of Professor A. Newton in "The Garefowl and its Historians,"[20] we are led to understand that this man lived until a short time prior to 1865. Major H. W. Fielden, when he visited

---

[13] Appendix to Supplement of "Montagu's Ornithological Dictionary," 1813.
[14] "The Garefowl and its Historians," Natural History Review, 1865, p. 473.
[15] Dr. Latham, "General History of Birds," vol. x. pp. 56, 57.
[16] "The Garefowl and its Historians," Natural History Review, 1865, p. 473.
[17] "Beskrivelse over Faerœerne," 1800, p. 254. "Landt."
[18] "Contributions to Ornithology," 1850, p. 115.
[19] "Reise nach Faro," pp. 198, 199.
[20] "Natural History Review," 1865, p. 476.

Faröe, saw a man named Jan Hansen, then eighty-one years of age, who told him that a Great Auk was caught on 1st July 1808.[21] If this man's statement be true, he must have had a wonderfully retentive memory to remember the date so exactly.

ICELAND.

Some of the skerries off the south-west coast of Iceland were, it is believed, the last breeding-places of the Great Auk. During earlier times the bird had a wider distribution around this coast, and for that reason we must refer separately to the different skerries it is said to have inhabited.

GEIRFUGLASKER, OFF BREIDAMERKURSANDR.

Professor W. Preyer, in his paper *"Ueber Plautus impennis,"* 1862, p. 25, states that E. Olafsson[22] mentions an island situated some geographical miles (probably fifteen or twenty English miles) off the Breidamerkursandr (Breidamerkur Sands), named the Geirfuglasker (*Anglice,* Garefowl Skerry), and if this is the case there can be little doubt it derives its name from the Great Auk. Olafsson says, " It gets its name from the Auk with the eight furrows on its beak ; " and though this is not quite true regarding the number of furrows on the bill of *Alca impennis,* which vary in number and are generally more numerous, what other bird would so closely answer the description, and be of sufficient importance to give its name to this rock ? There was only one bird known by the name of Geirfugl in Iceland so far as we have been able to discover, and that was the Great Auk. Whether this skerry was a breeding-place of the Garefowl, or only one of the islets it frequented, may be left an open question. Olafsson refers to this skerry as if its existence was only traditional. Professor Steenstrup says it is situated on the south coast of Iceland. It appears on the map in connection with his celebrated paper (see p. 2) situated nearly midway between the Westman Isles and Cape Reykjanes. Professor W. Preyer states the name is not now known even in Iceland. The skerry, if it appears on recent maps, is not given by the name of Geirfuglasker, and we are thus unable to identify it. It is therefore not marked on our chart.

---

[21] "Zoologist S.S.," p. 3280.
[22] E. Olafsson og B. Pálsson Reise igj Island. Soröe, 1772, p. 765.

GEIRFUGLASKER, EAST OF BREIDDALSVIK, EAST ICELAND.

Professor W. Preyer, on the same page of his paper as already quoted, refers to some rocks to the east of Breiddalsvik in East Iceland. He tells us they are mentioned by Olaus Olavius,[23] and are said to stretch a long way out to sea. A number are just visible above the ocean, but about six or seven German miles (about twenty-four or twenty-eight English miles) off the coast is a tolerably large rock of considerable circumference called the *Geirfuglasker*. To this rock in times past expeditions went about St. John's Day (near mid-summer) to catch Garefowls and seals. From the context this may have been a breeding-place of the Garefowl, and probably the date was chosen from the fact that the young birds would generally be hatched out by the end of June. We can hardly bring ourselves to think that it was with any desire to spare the young, and thus permit the perpetuation of the race, that this period of the year was chosen, as all the information to be had regarding these expeditions indicate that the fowlers slaughtered old and young indiscriminately. It is likely there were some holidays at this season, and as the weather might be expected to be usually better than at other periods of the year, this may account for the time chosen for visiting the skerry. As the Icelanders had only open boats, they required to make such expeditions in favourable weather.

There is on Olsen's large scale map of Iceland, about thirty miles from the coast, a skerry named Geirfuglasker, and this seems likely to be the islet referred to. We have marked this islet on our chart as a probable breeding-place. It has, however, by an unfortunate mistake, been named "Fuglasker."

Professor A. Newton, in his paper on Mr. J. Wolley's "Researches," Ibis, 1861, p. 374, says : "The most eastern Geirfuglasker is situated some thirty miles from the coast, off the island of Papey, and the entrance of Berufjorðr, about lat. 64, 35 N., and long. 26° W. (of Copenhagen), and is commonly known to Danish sailors as Hvalsbak (Whalesback)." "On making all inquiries we were able on our arrival at Reykjavik (*probably in* 1858), we could obtain no recent information respecting the eastern skerry, of which we had at starting entertained most hopes. It appeared also that of the travellers who in the last century had published accounts of their journeys in Iceland, Olafsen[24] and Olavius[25] only had alluded to

---

[23] German translation of Olaus Olavius' "Journey through Iceland," 1787, p. 313.

[24] "Reise igiennem Island," &c., af Eggert Olafsen (or Olafsson). Soröe, 1772, p. 750.

[25] "Œconomisk Reyse igiennem de nordvestlige, nordlige, og nordostlige Kanter af Island," ved Olaus Olavius, &c., Kjöbenhavn, 1780, ii. p. 547.

this isolated rock as a station for the bird, though another of them, the Faröese, Mohr, was in 1781 for no less than two months at Djupivogr, on the mainland opposite, engaged in the pursuit of natural history."[26]   Mr. Wolley then states that having made up his mind not to attempt the journey to this skerry, he engaged a native of the district, a gentleman named Magnusson, who went as his envoy, and reached Berufjorðr in the month of June, and then, taking a boat, he proceeded to the island, which he rowed round, close enough to satisfy himself that there were no Garefowls on it, but the unfavourable state of the weather prevented him landing.   On his return to Reykjavik during July, Mr. Magnusson informed Mr. Wolley that there were no traditions in that part of the country of the bird ever having been there.

GEIRFUGLASKER, WESTMANNEYAR, SOUTH ICELAND.

To the south of Iceland is situated a group of islands named the Westmanneyar (*Anglice*, Westman Islands), called after the monks of the early Celtic Church, who came from Ireland to worship God in peace in these remote solitudes, free from the distractions of the world.[27]   The most southerly of these islands is known by the name *Geirfuglasker*, and it was undoubtedly a breeding-place of the Great Auk.   It is probably nearly a century, and perhaps considerably more, since this island ceased to be a station for *Alca impennis*.   Professor W. Preyer, when he wrote in 1865, says, "The Great Auk bred here more than half a century ago."   This Geirfuglasker is said to comprise three principal masses of rock, and on one of these the Garefowl is said to have bred.   The locality of this skerry is indicated on our chart.

Respecting this skerry Professor A. Newton says : "We heard on all sides that it was yearly visited by people from the neighbouring islands, and though we were told that some fifteen years before a young bird had been obtained thence, it was quite certain that no Great Auks resorted thither now."   This probably refers to inquiries made by Mr. Wolley during 1858.   In a footnote Professor Newton remarks, "Of course it does not follow, even if the story be true, that this bird was bred there.   Faber states ('Prodromus der islandischen Ornithologie Kopenhagen,' 1822, p. 49), that he was on the Westman Islands in July and August 1821, and that a peasant there told him it was twenty years since a

---

[26] "Forsög til en Islandsk Naturhistorie, &c.," ved N. Mohr, Kiobenhavn, 1786, p. 383.
[27] "Dicuile Liber de Mensura Orbis Terrae." Ed. Valckenær. Paris, 1807.

Great Auk (and that the only one of the species he had ever seen) had occurred there. He adds, that this bird and its egg, upon which it was taken, remained a long time in a warehouse on one of the islands, but had vanished before his arrival. We may, with Professor Steenstrup (l.c. p. 76, note), infer from this that the Garefowl, even about the year 1800, was a great rarity in the neighbourhood." [28]

THE FUGLASKER (*Anglice*, BIRD SKERRIES) OFF CAPE REYKJANES. [29]

It was, however, the skerries to the south-west of Cape Reykjanes that afforded a last shelter to the persecuted Garefowls, and for that reason some details will be interesting. This region is one of volcanic activity, and numerous upheavals and subsidences have taken place. Islands have appeared and disappeared, and from active craters the sea has at times been covered with a thick layer of pumice. The phenomena attending these eruptions have been generally so appalling that mention is made of them in Icelandic history. Professor W. Preyer states that in the year 1210 A.D., there was an eruption off Cape Reykjanes; in 1219, there is a doubtful reference to another; but from 1222 till 1226 we are told that there were continual eruptions, and four great outbreaks during that time.

In 1237 occurred what is described as the seventh eruption, and in 1240 the eighth, when a number of small islands were formed, and were seen from the coast, but afterwards mostly disappeared, while others came to view at different places in the same neighbourhood. A long period of quiescence followed this period of activity; but in 1422 the ninth eruption took place, when there appeared above the sea high rocks of considerable circumference. The tenth eruption occurred in 1583; the eleventh, exactly two hundred years afterwards, in 1783; the twelfth in 1830, when the Geirfuglasker disappeared on which the Garefowls bred, and to this circumstance we shall presently refer. The most recent eruption took place at the end of July 1884, and a correspondent of one of the Edinburgh newspapers, writing from Reykjavik, refers to it as follows: " Towards the end of July volcanic disturbances occurred in the sea off Cape Reykjanes, the south-west point of Iceland, which culminated in the appearance above the waters of a new island. Beyond the occurrence of several earthquake shocks, by no means rare phenomena in that part of Iceland, there was little manifestation of what was taking place in the deep, until, on the 26th of July,

---

[28] Mr. J. Wolley's " Researches," *Ibis*, 1861, p. 379.
[29] *Reykjanes* translated into English means " smoke cape ; " *Reykjavik*, " smoke bay."

the light-keeper at Cape Reykjanes observed that a new islet had appeared about twenty miles off the coast. It lies about fourteen miles north-west of Eldey, or the Mealsack, a high cylindrical volcanic rock whitened at the top by the deposits of seafowl, whence the name Mealsack, which forms a conspicuous object familiar to all who have rounded Cape Reykjanes. The new island had at first the shape of a flattened cone rounded at the top, but a large part of one of the sides has since fallen down. More recent accounts throw considerable doubt on this story, and it is said that a vessel has passed over the spot where the island was seen without discovering any trace of it. On the other hand the light-keeper is said to maintain the truth of his story. It is possible he may have been deceived by mirage."

As far as we have been able to ascertain, the skerries south-west of Cape Reykjanes had occupied the position shown on the enlargement of that region given in connection with our chart, for at least some centuries prior to 1830. During that year the rock named Geirfuglasker, which was the innermost of the two outer skerries, disappeared beneath the waves; and as this rock was the principal breeding-place of the Garefowls, the event had a most important result, as it greatly hastened the extinction of the species, compelling the birds to seek a home on skerries that were easier of access from the mainland. The four skerries that are shown on our map are called collectively the Fuglasker (Fowl or Bird Skerries), but they may be divided into two groups, each consisting of two islets—the innermost of the groups being named the Eldeyjar (Fire Islands), comprising Eldey (Fire Island), where the last of the Garefowls was killed in 1844, and Eldeyjardrángr (Fire Island Cliff or Rock), a precipitous stack of rock which was too steep for the Garefowls to ascend, as they could not fly. The fowlers, who visited it long ago, must have found it impossible to climb, as Professor W. Preyer, quoting Eggert Olafsson (E. Olafsson og B. Pálsson Reise igj Island Soröe, 1772, l.c., pp. 896, 831), mentions that Eldey and Eldeyjardrángr are so steep that no one can mount them, although in former times there were rope-ladders on Eldey.

Professor Newton says: " Lying off Cape Reykjanes, the south-western point of Iceland, is a small chain of volcanic islets, commonly known as the Fuglasker, between which and the shore, notwithstanding that the water is deep, there runs a Röst (Roost), nearly always violent, and under certain conditions of wind and tide such as no boat can live in. That which is nearest the land, being about thirteen English miles distant, is called by Icelanders Eldey (Fire Island),

and by the Danish sailors Meel-sækken (the Mealsack), a name, indeed, well applied; for seen from one direction at least, its appearance is grotesquely like that of a monstrous half-filled bag of flour, the resemblance, too, being heightened by its prevailing whitish colour. Not very far from Eldey lies a small low rock, over which it seems that the sea sometimes breaks. This is known as Eldeyjardrángr (Eldey's Attendant)." [30]

As to these two last statements we may remark that though the waves may in times of storm break over this skerry, still it may be a stack of rock rising to a considerable height above the sea, though neither so high or large as Eldey.

The meaning of the name Eldeyjardrángr, as given by Professor Newton, we supposed was a mistake, as *drángr* means a *cliff* or *rock*. Having written to Professor Newton with reference to this matter, he replied on 16th March 1885, "I would, however, say that in Iceland its application is not confined to 'a stack of rock rising abruptly,' for the Eldeyjardrángr is a gently sloping low rock, over which the sea at times breaks." Professor Steenstrup, writing us on 13th April 1885 regarding this subject, says, "*Drángr* is certainly a rock or cliff, but it is principally used as a name for a cliff or rock near to the coast, near a larger island or skerry accompanying, as it were, the coast or the island as a follower or attendant. Hundreds of names are so supplied by the Islanders." It is quite possible that the wash of the Atlantic waves, aided by the volcanic disturbances that have occurred in this region, may have reduced Eldeyjardrángr from being a high rock to be now a low one, with a gently sloping surface; but we think there is not the slightest doubt that at the time this skerry got its name it was a high stack of rock. That it was as we have described it, is mentioned by Eggert Olafsson, who was in Iceland from 1752 to 1757, and again from 1760 to 1764. In describing the skerries off Reykjanes, he says: "Eldey lies a mile (German mile) distant from this promontory, and close beside it Eldeyjardrángr, *a high rock*. On these places dwell Auks and other sea and mountain birds, but they are so steep that now no one can mount them." [31] It would be interesting to have fuller information upon the present state of this skerry. If it should turn out that it is still a rock of considerable height, it need cause no surprise that the sea breaks

---

[30] Mr. J. Wolley's "Researches," *Ibis*, vol. iii., 1861, p. 379.

[31] "Ueber Plautus Impennis," von Wm. Preyer, Heidelberg, 1862, p. 28. Quotation from Eggert Olafsson's work, "Reise. igj Island," p. 896, 831. The Auks referred to could not be *Alca impennis*, but were probably *Alca torda*, or *Arctica alle*.

over it at times, as the Atlantic waves must sweep in upon that coast with tremendous force.

The outer group of skerries may be called the Geirfuglasker (Garefowl Skerries). The inner of the two skerries, forming this group, was the Geirfuglasker proper, on which the Garefowls bred, and which was submerged in 1830. Professor Newton says, " Some ten or fifteen miles further out (than Eldeyjardrángr) are the remains of the rock formerly known to Icelanders as the Geirfuglasker proper, and to Danes as Ladegaarden (the Barn-building), in former times the most considerable of the chain, but which, after a series of submarine disturbances beginning on the 6th or 7th of March 1830, and continuing at intervals for about a twelvemonth, disappeared completely below the surface, so that now no part of it is visible, though it is said that its situation is occasionally revealed by the breakers."[33]   This islet formed a suitable breeding-place for the bird, as part of its west side sloped down towards the sea, which made it easy for the Garefowls to get upon the rock.   The outermost skerry is named the Geirfugladrángr (Garefowls Cliff or Rock), and appears to have got its name to distinguish it from the Geirfuglasker proper, just in the same way that Eldeyjardrángr got its name to distinguish it from Eldey.   The name Geirfugladrángr would seem to indicate that it is a stack of rock and difficult to climb, and thus was unsuitable for the Garefowl to breed upon. Writing in 1861, Professor A. Newton says, " Further out again (to sea), perhaps some six-and-twenty English miles from Reykjanes, rises another tall stack, called by Icelanders Geirfugladrángr, and by Danish sailors Greenadeer-huen (the Grenadier's Cap)."[34]   " The only hope that exists of finding the Garefowl in this region is, that at the submergence off Reykjanes a colony of these birds existed at the Geirfugladrángr, or went there from the sunken rock."[35]   " It is about the same distance from it as Eldey, only much further from Iceland, and owing to its remoteness, and the dangerous surf that breaks upon its shore, has never been visited by any inhabitant of the mainland."[36]   Unfortunately, the learned Professor's hope has not been realised, and as other twenty years have come and gone without even a single trace of the living Garefowl being found, we fear it must be considered as extinct.

The Geirfugladrángr was once visited by a Count F. C. Raben, a Dane.

[33] Mr. J. Wolley's " Researches," *Ibis*, vol. iii., 1861, p. 380.
[34] Ibid.
[35] Ibid. p. 396.
[36] " The Garefowl and its Historians," Natural History Review, 1865, p. 479.

He went along with Faber when collecting materials for his fauna of Iceland, and accompanying them was a Danish botanist named Mörck; but of the party only Count Raben landed, and he narrowly escaped losing his life when getting off the skerry. During this voyage, which lasted from the 29th June till the 2d July 1821, they also visited the Geirfuglasker, now submerged. It is of importance to observe that when at the Geirfugladrángr the party did not notice any Gare-fowls, and as they should have been seen there at that season of the year if any frequented the rock, it throws considerable doubts upon the likelihood of its having ever been one of their breeding-places. Professor Newton mentions that " All these rocks have been long remarkable for the furious surf which boils round them except in the very calmest weather. Still more distant is a rock to which the names Eldeyja-bodi, or Blinde-fuglasker, have been applied by Icelanders. This is supposed to have risen from the sea in 1783, the year of the disastrous volcanic eruption in Skaptafells-sysla, and soon after to have sunk beneath the waves." [37] Professor Newton may have seen some of these islets from a distance, but unfortunately neither Mr. Wolley or himself succeeded in visiting them. He says : " In 1858 Mr. Wolley and I remained at Kyrkjuvogr, with two short intervals, from May 21st to July 14th. Our chief object was to reach not only Eldey, but the still more distant Geirfugladrángr, on which, probably, no man has set foot since the Danish Count, in 1821, with so much difficulty reached it. Boats and men were engaged, and stores for the trip laid in; but not a single opportunity occurred when a landing would have been practicable. I may say it was with heavy hearts we witnessed the season wearing away without giving us the wished-for chance." [38]

The Geirfuglasker, now submerged.

It was at the Geirfuglasker, now submerged, situated above twenty-five miles south-west of Reykjanes on the mainland, that there occurred during the latter period of its history the greatest slaughters of the Garefowl. Since the beginning of last century it appears that this Garefowl colony has been several times in danger of extermination, as expeditions went to it year after year about mid-summer, if the weather was propitious.

The colony became at times so diminished in numbers that for a series of years no expeditions went to it; but as soon as it was discovered that the Gare-

---

[37] Mr. J. Wolley's "Researches," *Ibis*, vol. iii., 1861, p. 380.   [38] Ibid. p. 394.

fowls had again increased, immediately efforts were made to kill them. If we may believe Anderson, who wrote in 1746,[39] the only place in Iceland at that time where the Garefowl were to be found was at this Geirfuglasker. In 1752, Horrebow, in reviewing Anderson's work, says,[40] "The Garefowl was at that time plentiful at this skerry." He also adds, "The fowlers, when they visited this rock, filled their boats with the eggs of the Garefowl." Eggert Olafsson, who was in Iceland from 1752 till 1757, and again from 1760 till 1764, writing in 1772, says,[41] "The Garefowl is found on one of the Westmanneyar (*Anglice*, Westman Islands), and also on a skerry off Reykjanes." In the public library at Reykjavik is preserved a short but beautifully written manuscript giving an account of the Geirfuglasker. This manuscript, from internal evidence, appears to have been written about the year 1760. It mentions the marvellous numbers of birds found upon the rock, and states that the " Garefowl is there not nearly so much as men suppose ; " "that the space he occupies cannot be reckoned at more than a six-teenth part of the skerry," "and this only at the two landing-places ; further upwards he does not betake himself, on account of his flightlessness." [42] N. Mohr, who visited Iceland in 1780–81, writing in 1786 ("Forsög til en islandsk Natur-historie," p. 29), refers to the statement of Horrebow, "that the fowlers in his time (1752) filled their boats with the eggs of the Garefowl," and says it is exag-gerated. There can be little doubt that Mohr had good reason to make this remark, for the female Garefowl only laid one egg each season.

From the beginning of the present century the principal descents that were made on this skerry have been well recorded. The first of these was perpetrated by the crew of a privateer named the *Salamine*, commanded by John Gilpin, but probably owned by Baron Hompesch, who was on board. This vessel in 1808 visited Faröe, and her crew plundered Thorshavn, where they found a man named Peter Hansen, whom they forced to proceed with them as pilot to Iceland. They arrived at Reykjavik on July 24th. and repeated their previous outrages, and on their way from Iceland visited the Geirfuglasker, where they remained a whole day killing many birds and treading down their eggs and young. They left here on the 8th August, and on their way south called at Faröe, where they landed Hansen.

---

[39] "Nachrichten von Island, Grönland und der Strasse Davis." Frankfurt u Leipzig, 1747, S. 54.

[40] "Tilforladeliga Efterretningar om Island," 1752, S. 49. As mistranslations occur in the English edition of Horrebow, and we have been unable to consult his original work, the reader must accept the statements attributed to him with caution.

[41] E. Olafsson og Pálsson Reise igj. Island. Soröe, 1772, I. C. S. 896, 831.

[42] Professor A. Newton on Mr. J. Wolley's "Researches," *Ibis*, vol. iii., 1861, p. 381.

The next catastrophe to this Garefowl colony was in the year 1813. The war between Britain and Denmark had resulted in the inhabitants of the Faröes being almost starved for want of supplies, and their governor, Major Lobner, sent the schooner *Faröe* of twelve guns to Iceland for food, placing it under the charge of Hansen, as he was already acquainted with the coast. When they arrived off the Geirfuglasker they were becalmed, and a boat having been lowered, its crew visited one of the skerries, on which they found an immense concourse of birds, among them being many Great Auks. They killed all that came within their reach, and after filling their boat, numbers were left lying dead, as they intended to return for them. But as the wind freshened, Hansen made sail for Reykjavik, where about a week later they arrived on the 29th July. They had then twenty-four Garefowl on board, besides numbers that had been salted down.[43]

It seems probable that this skerry on which these birds principally bred might have been their home to the present time, if its volcanic submergence had not compelled the colony that inhabited it to seek a home on another islet, nearer the shore, named Eldey, which they had not previously frequented, where they became a much easier prey to their inveterate foe, mankind. (See chart.) The volcanic disturbances that caused the Geirfuglasker, off Reykjanes, to disappear, occurred in 1830, beginning about the 6th or 7th of March, and a colony of the Garefowl shortly afterwards appeared at

ELDEY,

" A precipitous stack perpendicular nearly all round. The most lofty part has been variously estimated to be from 50 to 70 fathoms in height, but on the opposite side a shelf (generally known as the 'Underland') slopes up from the sea to a considerable elevation, until it is terminated abruptly by the steep cliff of the higher portion. At the foot of this inclined plane is the only landing-place ; and further up, out of reach of the waves, is the spot where the Garefowls had their home." [44] Professor W. Preyer, Jena, quoting Eggert Olafsson, who published in 1772 his work (E. Olafsson og B. Pálsson Reise igj. Island, Soröe, 1772, 1. c. pp. 831, 896) says : "In former times there were rope-ladders on Eldey, and there can still be seen large nails in the rocks where the ropes were fastened." [45]

---

43 Mr. J. Wolley's "Researches," *Ibis*, vol. iii., 1861, pp. 384–386.
44 Ibid. p. 391.
45 "Ueber Plautus Impennis," von. William Preyer. Heidelberg, 1862, p. 28.

It was at this skerry that the last pair of Great Auks were killed in 1844. Their intestines and other internal organs are now preserved in the Royal University Museum, Copenhagen, but what became of their skins, bones, and other remains appears to be unknown.[46] (See also Appendix, pp. 7 and 13 *notes*.)

The capture of these two birds was effected through the efforts of an expedition of fourteen men, led by Vilhjálmur Hákonarsson; but only three landed on the rock, and they at great risk, namely, Sigurðr Islefsson, Ketil Ketilson, and Jón Brandsson. Only two Great Auks were seen, and both were taken—Jón capturing the one, and Sigurðr the other. This event took place between the 2d and 5th June 1844. It appears that this expedition was undertaken at the instigation of Herr Carl Siemsen, who was anxious to obtain the specimens; and the day following the return of the boat to Kyrkjuvogr, Vilhjálmur started with the two dead Garefowls for Reykjavik. On his journey he met Christian Hansen (son of the Hansen before alluded to, who piloted the two vessels to the Geirfuglasker), and sold them to him for eighty rigsbank-dollars, or about £9. By Hansen they were passed on to Herr Möller, who was at that time the apothecary at Reykjavik.

It is from this last station that most of the skins and eggs now found in European collections have been obtained, and it is believed that during the fourteen years they frequented this rock at least sixty Garefowls were killed. When the Geirfuglasker sank, the colony of Garefowls was scattered, as a few individuals made their appearance at one or two points along the Coast of Iceland, where some were killed. Vilhjálmur Hákonarsson revisited Eldey in 1846, and again in 1860, but neither he nor any of his party could see any Garefowls.[47]

MAINLAND OF ICELAND.

There is a report that in 1814 seven Great Auks were killed at Latrabjarg.[48] Professor A. Newton on Mr. Wolley's "Researches" mentions that a man named Thorwalder Oddsson found a Great Auk on the shore at Selvogr about 1803 or 1805, and a few years later two were killed at Hellersknipa, between Skagen and Keblavik, probably about 1808 or 1810. Another is said to have been shot a few years later near the same spot; this, as well as the other two, was eaten.

---

[46] Robert Gray, Esq., in "Proceedings of Royal Society, Edinburgh," 1879-80, p. 679.

[47] Mr. J. Wolley's "Researches," *Ibis*, vol. iii., 1861, pp. 390, 393. Professor Newton spells the name of the person he calls the apothecary at Reykjavik "Müller." Professor Steenstrup says the name should be Möller, and that the only claim this person had to be called apothecary was that he prepared skins.

[48] "The Garefowl and its Historians," Natural History Review, 1865, p. 479.

In July 1821, near the same place, two birds were killed by a man named Jón Jónsson with a sprit or gaff, while they were sitting on a low rock. The skins were sold, but the bodies eaten. There is an unauthenticated account of one having been killed somewhere in South Iceland in 1818, but Etatsraad Reinhardt records the death of one in 1828.[49]

Professor W. Preyer states that twenty of these birds were killed at the island of Grimsey about the time of the submergence of the Geirfuglasker, off Reykjanes. This island is situated to the north of Iceland, and is intersected by the Arctic Circle.[50] This report, however, needs confirmation, as Mr. Proctor, who visited that island in 1837, and was weather-bound there for several weeks, appears never to have heard of this occurrence, as he would most likely have done if it had been correct. The following letter from Mr. Proctor to Mr. R. Champley of Scarborough is interesting : —

"University Museum, Durham,
February 28, 1861.

"Dear Sir,—Yours of the 25th came duly to hand, and would have been answered sooner had I been at home ; and in answer I beg to inform you we have the Great Auk in our Museum—but not the egg. We got the skin from Mr. Reid of Doncaster, I believe, about the year 1834 or 1835. The Rev. T. Gisborne bought the skin in Doncaster for £7 or £8, I believe, but where it was killed or taken I do not know. I was in Iceland in the year 1833, and made every inquiry, and sought for it, but never saw a single bird. I went to the northern part of Iceland in the year 1837 in search of it again, and travelled all through the northern parts as far as Grimsey Island, a small island forty or fifty miles north of the mainland of Iceland, but could not meet with it. I found the Little Auk breeding there. I never saw the bird alive. I never had any other skin than the one mentioned above, neither have I ever had the eggs. I have the eggs of the Little Auk, and a great many other eggs on hand.—I remain, your most obedient servant,

"W. Proctor."

It is as well to mention that there is another island named Grimsey, which is situated near the entrance to Steingrimsfjordr in the Huna Floi, North Iceland ; but Professor W. Preyer's statement is so clear that this cannot be the island he refers to, and we can only conclude that his information regarding the more northerly island of the same name is incorrect.

---

[49] Professor A. Newton on Mr. J. Wolley's "Researches," *Ibis*, vol. iii., 1861, pp. 384–389.

[50] "Ueber Plautus Impennis," von William Preyer, 1862, p. 23. Professor Steenstrup informs us, 13th April 1885, that he thinks the occurrence of these twenty Great Auks at Grimsey more than doubtful.

OTHER LOCALITIES.

The other localities at which the occurrence of the Garefowl has been reported since 1800 are as follows: It is mentioned that a specimen was picked up dead on Lundy Island, but this needs confirmation.[51]   In May 1834 two specimens were captured near the entrance to Waterford harbour, and one of these is now preserved as a stuffed skin in the Museum of Trinity College, Dublin.   The other was unfortunately destroyed through the ignorance of its captors.[52]   In February 1844 the Rev. Joseph Stopford communicated to Dr. Harvey of Cork that a Great Auk had been obtained on the long strand of Castle Freke in the west of County Cork, and that the bird had been water-soaked in a storm.   He did not give any date for this event, but it is generally understood that a number of years elapsed before he wrote that this bird was got.[53]

Two Garefowls are said to have been seen in Belfast Bay on 23d September 1845,[54] and this instance is worthy of note, as, if it is correct, it is a year later than the date at which the last Garefowls were killed on Eldey.   The observer on whose authority this statement is made was a Mr. H. Bell, a wildfowl shooter, who related that he saw two large birds the size of Great Northern Divers (which were well known to him), but with much smaller wings.   He imagined they might be young of that species, until he remarked that their heads and bills were much more clumsy than those of the *Colymbus*.   They kept almost constantly diving, and went to an extraordinary distance each time with great rapidity.

It is stated, but not on good authority, that early in this century several Garefowls were at different times seen or caught on the French side of the English Channel.[55]   In Denmark, where recently the remains of Garefowl have been found in ancient shell-mounds (kitchen-middens), there is only one instance of its occurrence on record, and that is made by Benicken, who informs us that one was shot about the year 1790 in Kiel Harbour (now German territory).   The appearance of this bird has been several times noted on the eastern side of the Cattegat. Professor Nilsson was assured by an old fisherman in Bohus län that in his youth he had seen the Garefowl on Tistlarna.   Another specimen is mentioned, on the authority of Dr. Œdman, to have been killed off Marstrand at the end of last cen-

[51] Dr. Edward Moore Charlesworth's "Magazine of Natural History," vol. i. p. 361.
[52] Thomson's "Birds of Ireland," vol. iii. p. 238.
[53] Ibid. p. 238.   [54] Ibid. p. 239.
[55] Degland, "Ornithologie Eur.," ii. p. 529; also in M. Hardy's "Catalogue des Oiseaux de la Seine-Inférieure."

tury. A dead bird is said to have been found near Frederiksstad in Norway during the winter of 1838.[56] In addition to the instances narrated two occurrences of the Great Auk are reported from inland situations in Britain, but one of these has been declared by the late Sir William Hooker,[57] on whose authority the statement was made, to have been a mistake,[58] and the other made by Fleming is evidently also an error.[59]

When at St. Kilda in June 1880, Mr. R. Scot Skirving met Mr. Mackenzie, the factor for Macleod of Macleod, and heard from him that in 1844 he saw shot at a place on the coast of the island of Skye what he now believed to be a Great Auk, though at the time the bird was killed all he knew was that it was a stranger to him, and having seen a Great Northern Diver only a short time before, he did not think it was one of them. Mr. Mackenzie stated that he was taking a walk along the shore when he met a man named Malcolm Macleod, who was out trying to get a shot at sea birds. In the course of their walk they saw a very large bird, which Macleod succeeded in shooting, and as it was some distance from the shore they had to get a boat to secure its body. They thought in the distance it was a Great Northern Diver, but when they got the dead bird they were at once struck with its remarkable appearance, as it differed from any bird they had ever seen. What the remarkable differences were do not appear very clear, as they did not observe the bill to be strikingly large ; and what attracted Mr. Mackenzie's attention principally were the large *claws*, which were, so far as he recollects, one and a quarter inches broad and not more that one thick. He got Macleod to give him the feet of the bird, which he says he kept until 1860, when they were lost during a flitting. He did not notice anything particular about the size of the wings. It was many years after the bird was shot before Mr. Mackenzie saw figures of the Great Auk, but on seeing the pictures he thought that the bird he saw killed in 1844 was the same in appearance. Mr. Scot Skirving, having obtained the address of Macleod, who resided in Greenock, wrote him, and he remembered the shooting of the bird when with Mr. Mackenzie quite well, and says he shot another of the same kind immediately after near the same place. The words he uses are : "I shot the one like to-day and the other like to-morrow." . . . "I never saw the like of them before or after.

[56] "The Garefowl and its Historians," Natural History Review, 1865, p. 469. Professor Steenstrup, 13th April 1885, says, "All the reported occurrences of the Great Auk in the Cattegat are insufficiently attested."
[57] *Ibis*, 1861, p. 398, *note*.
[58] "Linnæan Society Transactions," vol. xv. p. 61.
[59] "British Animals," p. 130.

They were about the size of a goose, but more graceful in the shape of the body. I have seen the picture of the Great Auk in the ' Encyclopædia,' and it reminds me of the birds. I cannot say I remember every particular about the make of the birds. They were dark on the coat, with white breasts and small wings. I think I shot them about forty years ago." Macleod wrote this information in 1880, so by his account the birds would be shot about 1840 and not 1844. As Macleod must have been accustomed to observe sea birds, it is unlikely he would have passed over unnoticed the remarkable beak of the Great Auk, if these had been such birds; and besides, when Mr. Mackenzie got the feet of the first specimen, Macleod took away the head as a trophy, so he had every opportunity of remarking any peculiarities if such had existed.

We candidly confess having great doubts as to the possibility of these birds having been Great Auks, quite apart from the very doubtful evidence. If we take 1840 as the date at which they were shot, there is a slender possibility that two specimens of *Alca impennis* might have found their way to the shores of Skye; but if Mr. Mackenzie is correct, and he seems to have little doubt of the date he gives, then it is most unlikely and against all probability, as the last pair were killed at the beginning of June 1844 on Eldey. It is rather curious that during the correspondence that has taken place between Mr. R. Scot Skirving, Mr. Mackenzie, and Macleod, the locality in Skye at which the birds were shot is not mentioned. It is easy to throw doubt on the identity of these birds with the Great Auk, but it is not so easy to say what other birds they could be except Great Northern Divers, and perhaps they were only these after all in some particular state of plumage with which Mr. Mackenzie and Macleod were quite unfamiliar.

Some ornithologists have indulged in the hope that in some hitherto unexplored part of the Northern Seas we would yet find the living Garefowl, but it appears to us that all hope of such a discovery has long since died out. For many years past part of the standing instructions to the naturalists who have accompanied the Arctic expeditions has been, " Look out for the Great Auk ; " but expedition after expedition has returned without any trace of the living bird having been found. The Europeans resident in Greenland are well aware of the value of its remains and the interest that attaches to its existence. They have been on the outlook for it during the last twenty-five or thirty years at least, and yet they have not a single occurrence to report. We are indebted to Mr. R. Champley for kindly sending us the following correspondence from the Arctic explorers, Sir F. Leopold M'Clintock, Captains John Rae and Allen Young :—

48 Hardwick Street, Dublin,
9th January, 1860.

Sir,—In reply to your note, I have to inform you that the Great Auk has not been met with by any of the modern Arctic expeditions.   I was told in South Greenland that some twenty-five years ago a young specimen was obtained, but am not at all certain of the fact. The resident Europeans are quite aware of the value attached by naturalists to that bird, so have kept a sharp look-out for it.   I have myself collected birds during my four Arctic voyages, all of which are now in the Museum of the Royal Dublin Society.   I am not aware of there being any new species amongst them.—I am, sir, yours faithfully,

F. L. M'Clintock.

R. Champley, Esq., Scarborough.

H.M.S. "Bulldog," Portsmouth,
21st November, 1860.

Sir,—Nothing has come to my knowledge respecting the Great Auk during my late voyage to Iceland, Greenland, and Labrador.   Captain Young will have quite equal and perhaps greater opportunities than I have had of ascertaining whether it still exists in any of those Northern seas.—I am, sir, faithfully yours,

F. Leopold M'Clintock.

R. Champley, Esq.

---

43 Hertford Street, Mayfair, London,
26th November, 1860.

Sir,—I regret that I have little or no information to give you about the Great Auk, although I questioned many persons in Iceland about this rare if not extinct bird.   An ineffectual search for them was made some time ago on an island or islands N.W. (*sic* S.W.) of Iceland, where they had previously been not uncommon, as bones are found there still — Believe me, your obedient servant,   John Rae.

R. Champley, Esq.

Steam Surveying Ship "Fox," Southampton,
November 30, 1860.

R. Champley, Esq., Scarborough.

Sir,—In reply to your question whether we saw a Great Auk, I can only say that to the best of my knowledge the bird has not been seen for many years upon the south coast of Greenland.—I am, sir, yours obediently,   Allen Young.

CHAPTER IV.

THE REMAINS OF THE GAREFOWL—INTRODUCTION TO THE SUBJECT— DISCOVERIES IN NORTH AMERICA.

WE shall now endeavour to trace out the localities at which remains of the Garefowl have been discovered. The positions in which such remains have been, and are most likely to be found, are the breeding-places of the bird, where numbers might die a natural death, or being killed by the early mariners and fishermen, their bodies were left lying uncared for. Plenty of their fellows in better condition than themselves being found to provision the ships, their bleached bones were scattered along the shore; but others soon becoming buried in the immense deposits of guano, were then frozen by the intense cold of the Newfoundland winter, which bound up everything within its iron grasp, so that even the intense heat of summer did not melt the soil beyond a depth of from two to three feet.[1] During a visit to Iceland in 1858, Professor A. Newton and Mr. Wolley obtained a few remains, which seemed to have become imbedded in turf, that had been removed from the sites of old kitchen-middens.[2] But it appears to us that by far the most interesting, from every point of view, are those remains that have been discovered in the shell-mounds of North America,[3] Denmark,[4] and Scotland,[5] along with those found in an ancient sea cave on the coast of the north-east of England.[6] Those discoveries all point to the existence of this bird at one time in districts where it has long been unknown, and associate it with the early inhabitants of those countries.

FUNK ISLAND.

It is to Herr P. Stuvitz, a naturalist sent out by the Norwegian Government during 1841 to inquire into the state of the Newfoundland cod-fisheries, that we

---

[1] "Letter of Bishop of Newfoundland," Transactions of Nova-Scotian Institute of Natural Science, vol. i. part 3.

[2] Mr. J. Wolley's "Researches," *Ibis*, vol. iii., 1861, pp. 394–396.

[3] "American Naturalist," vol. i. pp. 374–578. J. Wyman.

[4] "Oversigt over Videnskab." Selskabs Fordhanlinger, 1855, S. 13–20, p. 385.

[5] "Prehistoric Remains of Caithness." Samuel Laing, M.P., 1866; and "Linnean Society's Journal, Zoology," vol. xvi. pp. 479–487.

[6] "Natural Hist. Trans. of Northumberland, Durham, and Newcastle-on-Tyne," vol. vii. part 2, 1880, pp. 361–364.

owe the first announcement of the discovery of remains of the Garefowl.[7] In his report he mentioned that immense numbers of a bird called the Penguin used to frequent the banks, and bred upon the islands off the coast.   His Government, understanding that the Penguin was a bird confined to the southern hemisphere, were inclined to doubt his information.   This led Stuvitz to visit Funk Island, and he there obtained a quantity of remains, which he sent home.   On the arrival of the remains in Europe they were discovered to have belonged to the Garefowl, and have proved most useful in identifying bones subsequently found in other localities.[8]

At the end of June 1841, Stuvitz was at St. John's, which he left on the 30th of that month, arriving at midday on the 31st at Funk Island,[9] which is 170 miles north of St. John's, and about 36 miles north-east-by-east from Cape Freels, the north headland of Bonavista Bay.[10]   There he found large quantities of Garefowl bones lying upon the shore, and the remains of compounds into which the birds had been driven to be slaughtered.[11]

For nearly twenty years we hear nothing more about the remains on Funk Island, until Professor A. Newton, of Cambridge, believing the Garefowl to be probably extinct, realised the value of the remains that might be still obtainable. He wrote to numerous parties in Nova Scotia, in the hope of interesting them, but without effect, until he received the promise of assistance from the Rev. Reginald M. Johnson.   In 1863 that gentleman made a journey to Funk Island himself, and the success that attended his efforts was beyond his expectations, as his researches resulted in the recovery of a mummy of the Garefowl and some bones. Mr. Johnson having communicated with the Bishop of Newfoundland, that gentleman wrote Professor Newton a letter, which was received on 7th November 1863, and the mummy, arriving about the same time, was submitted to a meeting of the Zoological Society, London, held on the 10th of that month.[12]

During the year 1864 other three mummies were dug out at Funk Island

---

[7] "The Garefowl and its Historians," Natural History Review, 1865, p. 484.

[8] "Videnskabelige Meddelelser," 1855, Nr. 3-7, p. 34.

[9] Ibid. pp. 63, 64.

[10] "Annals of Natural History," third series, part 14.   "Proceedings of the Zoological Society," Nov. 10, 1863.

[11] "Videnskabelige Meddelelser," 1855, Nr. 3-7, p. 65 (separate edition, pp. 33).

[12] "Annals of Natural History," third series, part 14, p. 435. " Proceedings of Zoological Society," 10th Nov. 1863.   This paper is from the pen of Professor A. Newton, and the writer says (p. 437) :—" It appears that the Colonial Government have recently conceded to a Mr. Glindon the privilege of removing the soil from Funk Island ; for this soil, being highly charged with organic matter, is consequently valuable as manure when imported to Boston and other places in North America."   Professor A. Newton then informs us that the workmen appear to have done their work very effectually, " for I hear that they brought away many puncheons of bones and other remains—of course not all necessarily Penguins."

from at least four feet below the surface, and from under ice which never melts. The Bishop of Newfoundland, in writing to the Nova Scotian Institute of Natural Science, telling of the disposal of these specimens, says—" One is sent to Mr. Newton, another to Agassiz, and one to yourselves ; and it is better than that sent to Mr. Newton, and possibly better than that sent to Agassiz, which I have not seen." [13]

On the 20th July 1874 Funk Island was visited by Mr. (now Professor) John Milne,[14] who wrote an account of what he saw for the *Field* newspaper, and afterwards published it in a separate form, under the title of " Relics of the Great Auk on Funk Island." He describes the island as having the appearance, at a distance of half a mile, of a smooth-bottomed upturned saucer, slightly elongated into an ellipsoidal form towards its north-eastern extremity, from which end it sloped more gradually up from the sea than it did from its opposite end. The island has a few boulders, and the rough stonework remains of several compounds on its surface. The landing is rather difficult. The number of sea-fowl very great, which rise in a shrieking and wailing cloud above the visitors. He says—

" Having a strong wish to secure some relics of this bird, and my time for their discovery being limited to less than an hour, it was with considerable excitement that I rushed from point to point and overturned the turf. At nearly every trial bones were found, but there was nothing that could be identified as ever having belonged to the bird for which I searched. At the eleventh hour the tide turned, and in a grassy hollow, between two huge boulders, on the lifting of the first sod I recognised the alcine beak. That rare element called luck was in operation. In less than half an hour specimens, indicating the pre-existence of at least fifty of these birds, were exhumed. The bones were found only from one to two feet below the surface, and in places even projected through the soil into the underground habitations of the puffins. With the exception of one small tibia, and two or three tips of long thin beaks, probably those of the tern, all the bones were those of the Great Auk. . . .

" In several cases, whilst exhuming the skeletons, I noticed the vertebræ followed each other successively, and were evidently in the same position which they occupied when in the live bird. This is in part confirmed by one curious case, where the rootlet of some plant had grown through the neural canal, and expanded so as to fix the vertebræ in position. This, together with the fact that there remains no evidence of cuts or blows, leads to the supposition that these

---

[13] " Transactions of Nova-Scotian Institute of Natural Science," vol. i. part 3.
[14] Is now Professor in the School of Engineering in Tokio, Japan.

birds may have died peacefully.   Nevertheless, it may be that they were the
remains of some great slaughter, when the birds had been killed, parboiled, and
despoiled only of their feathers, after which they were thrown in a heap, such as
the one I have just described." [15]

Among the bones discovered, one fragment alone showed signs of having been
burnt.   But a feature of the other remains that struck Professor Milne was that
some of the bones varied much in size ; and, from our own experience, we think
it quite possible that certain bones may belong to the *Alca torda* or razorbill, as
the humeri are difficult to distinguish from those of the *Alca impennis*, except
by the size.   Professor Owen refers to this resemblance. [16]

In a recent work on Newfoundland it is mentioned that the Penguin or
Great Auk has now entirely disappeared from that coast.   Incredible numbers
were killed at Funk Island, their flesh being savoury food, and their feathers
valuable.   There being no wood on the island, heaps of them were burnt as fuel,
in order to warm water in which others were steeped, with a view to the soften-
ing of their skins, and the consequently easier extraction of their feathers.   The
merchants at Bonavista at one time used to sell these birds to the poor people by
the hundredweight instead of pork. [17]

Another writer [18] mentions that the Great Auk " was not rare " in certain
parts of Newfoundland within remembrance of the present generation ; but we
think this statement is made on mistaken authority.

In 1867 a Mr. Wyman [19] found perfect limb-bones of the bird in shell mounds
near Portland, Maine, and in Massachusetts, U.S.   The remains are said to
represent parts of at least seven individuals. [20]   This discovery should encourage
American archæologists to make further search in similar shell deposits near their
coasts, for we feel sure their labours would be amply repaid.

---

[15] " Relics of the Great Auk on Funk Island."   Mr. John Milne, 1874.   Also in *Field* newspaper, 27th
March, 3d and 10th April 1875.

[16] " Zoological Transactions," vol. v. p. 330.   Professor Owen's Description of Skeleton of *Alca
impennis*, L.

[17] " Newfoundland as it Was, and as it Is in 1877."   By the Rev. Philip Tocque. M.A.   London and
Toronto, 1878.   Mentioned by R. Gray, Esq., "Proceedings Royal Society," Edinburgh, 1879–80, p. 677. '

[18] " Field and Forest Rambles."   A. Leith Adams, p. 36.

[19] Professor Steenstrup, in a letter dated 15th March 1885, has been kind enough to inform us that the
Mr. Wyman referred to is Professor Jeffries Wyman, the celebrated anatomist.

[20] " American Naturalist," vol. i. pp. 574–578.   Also in a paper by Professor James Orton.   The copy
of this paper, which we have seen, has evidently been cut out of some scientific magazine, or the proceedings
of a society ; but we do not know its source further than that it is evidently from the United States.   It
treats of the American remains, and refers specially to the skin preserved in the Smithsonian Institution,
Washington.   In a footnote to page 540 the author says, that in New England bones of the species (Great
Auk) have been discovered in shell-heaps at Marblehead, Eaglehill in Ipswich, and Plumb Island.

CHAPTER V.

THE REMAINS OF THE GAREFOWL IN DENMARK AND ICELAND.

HAVING now treated of the American locality, we shall turn to the European region, and endeavour to give a short account of the remains that have been there discovered.

DENMARK.

We believe it is to Professor Steenstrup of Copenhagen that the credit attaches of having identified the first bones of the Garefowl known to have been found in Europe, which were discovered at Meilgaard in Jutland. As we are not aware that any of his papers have been even partially published in English, we have obtained the permission of Professor Steenstrup to give some translations of parts of them, kindly furnished us by a friend. The first extract we give is the beginning of what may be considered the most valuable paper that has been written on the Great Auk, and, it was the discovery of the first bones of the *Alca impennis*, L., found in Denmark, that led Professor Steenstrup to study the history of the bird. He says [1] —

" In the investigation of the kitchen-midden of the primeval people (Oversigt over Videnskab, Selskabs, Forhandlinger, 1855, S. 13–20 og 385–388), among other remains were found some traces of two of the larger birds not now found here, viz., the Capercaillie (*Tetrao urogallus*, L.), and of a larger bird of the Auk tribe, which must be regarded as the as good as extirpated Garefowl. Since the last-named bird has not been found in the last few decenniums—no, not even in this century—breeding upon any place farther south than the Garefowl Rocks, that lie some few miles from the south coast of Iceland; and since the further appearance of scattered specimens, driven towards the north or western coasts of Europe, belong to the class of the greatest rarities, the proof of the existence of many specimens of the Garefowl in this heap must naturally be very surprising, since

---

[1] " Et Bidrag til Geirfuglens, Alca impennis, Lin., Naturhistorie og saerligt til Kundskaben om dens tidligere Udbredningskreds af Jap. Steenstrup, Professor." Videnskabelige Meddelelser, 1855. Nos. 3–7, with a Plate and Map.

that must indicate that this bird was three or four thousand years ago found down the Cattegat.

"The more unexpected this discovery was, the more important was it for me to be able to place beyond doubt the explanation of the discovered bones. This was all the more difficult, as no skeleton of this rare bird existed either in our own museum, or, so far as is known, in any other museums.[2] But as I, on the one hand, found perfect agreement between the discovered bones and the corresponding bones of all the lesser European birds of the Auk tribe, and on the other hand found certain peculiarities that distinguished the former from the latter, I could scarcely go wrong in declaring the bird to which the bones belonged to be, in the first place, a bird of the Auk tribe; in the second place, an Auk of the size of a goose; and lastly, an Auk in the highest degree fitted for swimming and diving, but utterly unsuited for flight,—a state of matters only applying among all known species of this family to the *Alca impennis* of Linnæus. In this opinion I was quite confirmed by a remarkable combination of circumstances. Among a little circle of Scandinavian naturalists it was known that the Norwegian naturalist, P. Stuvitz, whom his Government had sent out on account of the fisheries to Newfoundland and the adjacent parts of the North American continent, had sent home some bones of birds from a little island off the coast either of Labrador or Newfoundland. These bones were found in large heaps on the shore, and after their arrival in this country they were declared to be the bones of the Garefowl. This could be asserted of them all the more certainly as there were found among them, in addition to all the essential bones of the skeleton, a not inconsiderable number of crania, and these crania agreed in every respect with those which one had from the few stuffed specimens. Some of these bones, sent by Stuvitz, had luckily been presented to the Zootomical Museum of the University, and, moreover, some of them belonged to the same parts of the skeleton as the bones now under discussion that had been found in the primitive kitchen-middens.[3] On laying the two sets side by side, one could not doubt that it was birds of the same species that had been eaten in both places, and thus the bones from our kitchen-middens found the best materials for their perfectly certain

---

[2] For list of skeletons, see p. 82. These were not recorded at the time Professor J. Steenstrup wrote his paper, so were quite unknown to him.

[3] In a letter dated 15th March 1885 Professor Steenstrup informs us as follows regarding these bones :— "Presented to the Zootomical Museum of the Royal University here. The director of this museum was at that time Professor Eschricht. After his death the contents of this museum went partly to the Zoological and partly to the Physiological Museum."

explanation in bones from similar refuse heaps on the east coast of North America, which up to that time had not received the attention which they deserved."

The foregoing gives a very accurate account of the obstacles that had to be overcome by the now venerable scientist, and every person interested in the history of the Garefowl must feel under a debt of gratitude to him for his labours and be encouraged by the success of his investigations when surrounded by apparently insurmountable difficulties.

The bones from Meilgaard were three in number—viz., two right humeri and a radius from the right side of the bird, and were discovered in an ancient kitchen-midden. Two of these bones were figured (a humerus and a radius, two views of each) along with three bones of the Capercaillie, and form the principal plate to the elaborate paper on the Great Auk published by Professor J. Steenstrup, and from which we have just given a short quotation. This plate has been reproduced with even greater excellence than at first, and appears in connection with another Danish publication issued during 1855,[4] and in which the following description of the bones is given along with the writer's comments. He says—" I have recognised the remains of the Great Auk or Garefowl, *Alca impennis*, L, in several bones from Meilgaard. That this large northern bird, the only bird in Europe which on account of the smallness of its wings is not in a condition to fly, should occur among the remains of the meals of our primeval people must certainly appear in the highest degree striking, but it receives corroboration from the following bones found there :—

" I. *A right humerus*, agreeing entirely in form with the humerus of an *Alca torda* or razorbill, but in size double of it, 4 inches (Danish) long, 7 lines (Danish) broad, and, since the bone formation in this group of birds is so significant, certainly pointing to an Auk in size like a goose. The bone belongs fortunately to the few that have not been injured by the gnawing of dogs, but it bears on its surface one or two sharp scratches, marks of the knives of the primeval inhabitants. It belongs to an old bird.

" II. *Another humerus* from the same side, and consequently belonging to another individual. It is not nearly so much compressed as the foregoing, and thus belongs perhaps to a somewhat younger bird ; both its ends are bitten off, and

---

[4] Undersogelser i geologisk—antiqvarisk Retning af G. Forchhammer, Etatsraad og Professor, J. Steenstrup, Professor, og J. Worsaae, Professor. Kjöbenhavn, 1855. Translated—Researches in Geology and Antiquities, by Privy Councillor and Professor G. Forchhammer, Professor J. Steenstrup, and Professor J. Worsaae. Copenhagen, 1855, pp. 165–171. (Reprint from Proceedings of Royal Danish Society of Sciences.)

bear marks of the edges of teeth.   The cavity inside is less than in the *Alca torda*, and points to the conclusion that the bird was not able to fly.[5]

"III. *The radius* from the right side of an old bird, slightly injured at the ends.   Its peculiar short compressed form is entirely different from that of the *Alca torda*, and shows conclusively that the forearm has been short, and that the wing has not been adapted for flight, as is the case in the *Alca impennis*, L.

" Now, unless we suppose that these bones belong to some extinct and hitherto unknown species of Auk, which has had not only the same size but the same proportion in its wing as the Great Auk, we can only refer them to that now as good as extinct bird.   On placing the bones above mentioned alongside of the corresponding bones of the Great Auk and comparing them, I found such a complete agreement that I cannot entertain the slightest doubt that they belong to that species.   As, however, it must be acknowledged that no skeleton of this bird exists in our museums, and indeed will hardly be found in any museum in Europe,[6] I shall add that that comparison was conducted in the case of the *radius* by means of the perfect forearm and hand (wing and wing extremities), which were taken for that purpose out of the stuffed specimen of this rare bird belonging to the University ; whilst the *two humeri* were compared with some loose bones which Professor Eschricht at the Naturalist's Congress at Christiania got for the Zootomical Museum of the University,[7] and which the deceased Norwegian naturalist, P. Stuvitz, had collected in his time on an island near the coast of Labrador (*sic*, should be Newfoundland) ; it so happening by good chance that among them there were found two humeri, and these too belonging to the same side of the bird as the humeri in question.

" The unexpected appearance of this bird's bones in a kitchen-midden in the innermost part of the Cattegat cannot but call forth speculations as to the cause of its being found there, seeing that the Great Auk now lives so far from our coasts, and has such a limited area of diffusion.   It is certainly possible that its

---

[5] We do not give a reproduction of the plate in connection with Prof. J. Steenstrup's valuable paper, as we give illustrations of humeri found in Scotland.   See p. 44 and Plate p. 86.

[6] At the time the above was written, the learned Professor was evidently unaware of some of the skeletons of the *Alca impennis*, L., preserved in European Museum (see p. 82).   Professor Steenstrup writes us on 15th March 1885, and referring to the note says, "I think that the respective museums did not know themselves that they possessed the inquired for skeletons.   In Paris, at least, until 1859, the skeleton could not be found.

[7] Professor Steenstrup informs us, in a letter dated 15th March 1885, that "Zootomical Museum " is the correct term.   In the original from which the translation is made, the bones are said to have been got for the Anatomico-Physiological Museum, and it is probable they are now preserved there.   (For explanation, see note, p. 32.)

appearance there is entirely due to chance, and that the individuals from which the bones proceeded may have been isolated birds driven there by stress of weather, but still that is not very probable.   For it is to be remembered that the bones belong certainly to two, and perhaps to three individuals; that consideration for one thing diminishes the probability of the bird's appearance there being merely casual. Then, moreover, the remains of these individuals turned up in the very small part of the midden which has been examined this year (1854), and there is certainly no ground for assuming that the part hitherto unexamined, and that is by far the larger part, will be destitute of a due share of Auk bones in proportion to other bones.   Such at least is our experience with regard to the other rarer animal remains preserved in these refuse heaps.   The bones of the beaver, the marten, and the wild-cat have continued, on the whole, to turn up in essentially a uniform ratio to the other numerous bones as the excavations are gradually carried on.

"But should we agree to the probable supposition that there really appear here remains of the Great Auk representing not a few individuals, we must also take as granted a more regular appearance of this bird upon our shores in former times,[8] nay more, since that would not agree well with the bird's present limited diffusion over the northern seas, we must assume a wider area for it altogether in past ages, for the only spot on the seas of northern Europe where this bird is known to have had during this century a place of resort at all fixed are the little volcanic rocks to the south of Iceland,[9] called the Geirfuglasker (Great Auk Skerries), on which indeed small bands of these birds seem to have had something like a permanent abode; but yet these bands were so limited in number that a casual capture of from twenty to thirty of them, such as took place for example during 1830–31, and in the commencement of the century in 1813, appears to have left its impression for a long time thereafter in the diminished number of the birds observed or caught by later travellers.

"Outside of these rocks an isolated bird has now and then, and after an interval of some or many years, been seen on different spots of the southern shores of Iceland; but everything shows that such have been only casual visitants.   During last century, on the other hand, it seems also to have lived,

---

[8] More recent discoveries during excavations in Denmark, and to which we refer in the pages immediately following, show that Professor Steenstrup's supposition is correct, and that the *Alca impennis*, L., was no casual visitant to the shores of the Cattegat in early times.

[9] The only one of the Geirfuglasker on which the Great Auk bred in recent times was situated off Cape Reykjanes, on the south-west of Iceland.   This skerry was submerged during a volcanic disturbance in 1830 (see p. 18), when the Great Auk found a new breeding-place on Eldey, p. 20.

although only in exceedingly few and rare individuals, on the Färöe Islands (Mohr. Landt.), perhaps also on the rocks off the coast of Söndmör (Ström), whilst already at the beginning of that century it had become a casual visitant, which was seen only now and then after intervals of many years, at St. Kilda, the westmost of the Scottish islands (Macaulay), where, however, it had regularly bred, on to near the end of the seventeenth century (Martin).

"On the American side of the North Atlantic Ocean the Great Auk does not seem to have had any better fate, for the older accounts (those of Egede, Crantz, Glahn, Fabricius) show us clearly enough that the Great Auk was seen regularly a hundred years ago, although extremely rare, off the coasts of Southern Greenland, and that it bred there at least now and then (Fabricius); but these few remains had, as it seems, altogether disappeared by the beginning of this century (Hollboll and Reinhardt).

"Its appearance on the Labrador coast is more than doubtful. Now, as we know no bird to which nature has given such a limited diffusion and an existence in so few individuals, we are bound, after the analogy of all other higher animals, and especially of those more nearly allied, to assume that the Great Auk has in times past had a much wider area of dispersion ; and as we have been able in the last two or three centuries to see it disappear more and more from the more southerly zones of its area, it is reasonable to suppose that this limitation in its diffusion had begun in earlier centuries, and that in these early times the bird went much further down along the coasts of England (*sic*, he means Britain) and Scandinavia, but was gradually forced to disappear from them in consequence of the persecutions of man. For, being unable to fly, it was entirely in man's power whenever it crept up upon the lower rocks and cliffs of the shore, as, of course, must have been the case during hatching time ; and even if the old bird should escape man by seeking the water betimes and abandoning to him its egg or its downy chicken, still the species would suffer then a heavier loss than would any of the other birds inhabiting the sea cliffs, inasmuch as these in general attempt to lay a second or a third set of eggs, which the Great Auk never does. From such a southern breeding-zone it might easily be conceived that the bird also could reach the Cattegat more regularly during the colder months by swimming—if it should not be preferred to suppose the birds lived here also through the summer months, which I do not think at all improbable. It is difficult to see how it should not thrive here equally well with the many birds allied to it, the razorbill, for example. Naumann thinks that the history of the puffin presents a similar phenomenon, for in former times that bird inhabited the cliffs of Heligoland in a large colony, but has now diminished to a very small one, and when that little colony has disappeared, as will be the case, it is to be feared, in a short time, then the zone within which the puffin breeds will have its limit shifted all at once several degrees to the north and west.

" Writers often express themselves regarding the Great Auk as if they assumed that its individuals withdrew northwards in consequence of persecutions to more inaccessible places, and as if they ought to be found in large numbers along the shores of Spitzbergen and North America; but to judge from what we have learned up to this time (1855), there would seem to be little foundation for any such assumption. Stuvitz came upon bones of it in large numbers on the little island of *Fogo* (*sic,* should be Funk Island), off the shores of Labrador (*sic,* should be Newfoundland), but these bones belonged solely to birds that had been eaten; living ones do not seem to exist on these coasts. Did they exist there they would be seen now and then, specimens would occasionally find their way to Europe, and the bird would not remain the rare and costly thing it is.

"In conclusion, it is a noteworthy coincidence that the bones of this bird, found in the kitchen-middens of the primitive inhabitants of Denmark, should find their best elucidation in bones found under similar circumstances in Labrador (*sic,* should be Newfoundland)."

The foregoing are the conclusions formed in the mind of the learned Professor at the time of the discovery of the first remains of the Great Auk in Denmark, but since that time his opinions have received remarkable confirmation; and during the summer of the following year (1856), a second discovery was made in a kitchen-midden at Havelse, situated at the southern part of the Issefiord in Seeland, which led Professor Steenstrup to write another paper on[10] the *Alca impennis,* L. Of that paper we are enabled to give the following translation :—

" From the kitchen-midden of the primeval inhabitants at Havelse, I have during the present summer (1856), got for the museum a number of bones, partly collected by myself and partly by Herr Feddersen, who by the work of collecting has more than once aided my palæological researches. The most important of the contributions thereby made to the investigations of our pre-historic fauna

---

[10] Undersogelser i geologisk—antiqvarisk Retning af G. Forchhammer, J. Steenstrup, og J. Worsaae, Professors. Kjobenhavn, 1856. Translated—Researches in Geology and Antiquities, by Professors G. Forchhammer, J. Steenstrup, and J. Worsaae. Copenhagen, 1856, pp. 185-88.

is undoubtedly a *humerus* dug up by Herr Feddersen, belonging to a bird of the so-called black family (*Svartfugle*),[11] and of that family in particular to an Auk. That bone has at the same time belonged to an individual that must have been larger than the largest specimen of our common Auk, the razor-bill (*Alca torda*, L.), whilst it is considerably smaller than the corresponding bone of the Great Auk (*Alca impennis*, L.), as will be seen if the reader will compare the woodcuts on the next page with figure 4 in my last paper (see Professor Steenstrup's original papers, or compare reproduction of woodcut, p. 38, with figs. 10, 11, 12, 15, plate, p. 86), which represents in exactly its natural size the same bone of the Great Auk, and that too from the same side. Besides the size of the bone, we must also notice its very small and much compressed inner cavity, and its exceedingly thick osseous walls. The smallness of the cavity shows that the bone could not have come from any flying bird, and, taken along with the thickness of the

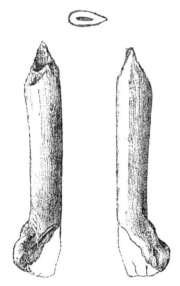

Reproduction of figure of a *humerus* dug up by Herr Feddersen in 1856.

osseous walls, shows also that that want of capacity to fly is not to be explained by the hypothesis that the bird was a young one with imperfectly developed wings, but rather by supposing that it was an old bird of a kind altogether unsuited for flight, as is the case with the *Alca impennis*, L. If, contrary to what now seems most natural, an increased supply of Great Auk bones should prove that the bone rather belonged to a young bird, and this may happen, for there does not seem to be at present in the museums of Europe a single perfect skeleton of the bird,[12] much less bones of birds of different ages, then we have proof not only that the bird was found in prehistoric times on our islands and their fjords, but also that it lived there during its breeding season. But if the bone belongs, as everything seems to indicate, to an old bird, then it shows either that the *Alca impennis*, L., must have presented an unusual diversity of size, or else that it must

---

[11] "Svartfugle"=the family of *Alcaceæ*, containing the genera *Uria, Alia, Mergulus, Mormon*, named black birds because their colour is so dark, contrasting with the white or whitish colour of the different species of *Larus* and *Fulmarus*, nesting on the same places, or seen in their neighbourhood.

[12] See note, p. 34.

have had co-existent with it a smaller species likewise unsuited for flight, which has disappeared and been extirpated along with it. This latter appears to me the more probable theory. Be that as it may, however, this bone certainly raises interesting questions, materials for whose solution it will be important to gather." [13]

Professor Steenstrup then goes on to state his opinions as to the area over which the Great Auk has been distributed. He refers to the great numbers of the birds on the islands off the coasts of North America during historic times, mentions Funk Island as its last stronghold in that region, and states that the mistake he made in his former paper [14] regarding the name of this island, when referring to the remains obtained by Herr Stuvitz, arose from the tickets attached to the bones bearing the name of Fogo Island.

In answer to our inquiries for information regarding the further discoveries of Great Auk remains in Denmark, Professor Steenstrup writes as follows, under date 8th April 1882 : " The other localities where I have found, or from where I have got the bones of the Garefowl, are the kitchen-middens at Fannerup, not far from Meilgaard, and at Gudumlund, some English miles south of the Limfjord, consequently also in Jutland, and here in Seeland from Sölager kitchen-midden situated at the northern part of the Issefjord."

Some of the Kjökkenmöddings or kitchen-middens have long been known in various parts of Denmark, and at one time were supposed to be raised beaches, because they were found scattered along the sea coast, especially on the slopes or banks of the numerous fjords which now or formerly intersected the country. At length it was discovered that these supposed raised beaches were really artificial, and contained the remains of a prehistoric population and fauna that belonged to

---

[13] In connection with what Professor Steenstrup mentions, we would refer the reader to a remark made by Mr. Eyton, who became the possessor of one of the Great Auk skeletons constructed from bones brought from Funk Island (see p. 100). In writing to Professor William Blasius, who quotes his letter in his work ("Ueberreste Von Alca impennis," p. 133), he says : " This skeleton was so different in appearance from that to be found in Newton's collection and figured by Owen, that he could almost suppose there had been two species of Great Auk." Professor Blasius, after quoting this statement, remarks : "To judge from the figures, Eyton's skeleton appears to have belonged to a small female bird, whereby probably the diversity can be explained." Professor Steenstrup, in a letter to us dated 4th February 1885, says : " ' Unknown variety of Great Auk,'—so I then said ; but since that time the many bones of *Alca impennis*, brought from a small island near the coast of Newfoundland (*Funk Island*), have shown us a rather great variation, I think not less a quite *individual* than a sexual variation." Professor Newton, in his paper on Mr. J. Wolley's " Researches," *Ibis*, 1861, p. 395, makes the following statement regarding the bones found in Iceland : " Among the specimens we collected there are several in which certain differences, probably the result of age or sex, are observable."

[14] See page 37 of this vol.

the Stone Age.　As might be expected, the Kjökkenmöddings vary greatly in size and appearance.　One of the largest and earliest explored is that of Meilgaard, situated about two miles from the sea in a beautiful beach forest called "Aigholm Wood," between which and the sea are high dunes of drifting sand, through which the tops of trees are sometimes seen protruding.　This shell mound covers an oblong space of about 340 feet in length and 120 in breadth, with a deposit of a maximum thickness of 10 feet.　The Kjökkenmödding at Fannerup is now about ten miles from the sea, situated on the border of a flat district which within historic times has been an arm of the sea, but afterwards became a fresh-water lake, and is now to a certain extent dry land.　The one at Gudumlund was situated on a southern expansion or bay of the Limfiord, but at present is separated from the sea by an extensive peat bog.[15]

We have been unable to ascertain the date of the discovery of the Fannerup middens, and also the number of bones of the *Alca impennis*, L., obtained at this locality.　The excavations at Gudumlund and Sölager were continued during 1873, and resulted in the finding of bones of three individuals at each place.[16]　The work at both places was conducted by the director of the Zoological Museum and his friends, and the expenses of the excavations at Sölager were borne by the Royal Danish Society of Sciences.　The result of these investigations at the dwelling-places of the ancient inhabitants of the coasts of Denmark has been

---

[15] "Danish Kjökkenmöddings, their Facts and Inferences." By Robert Munro, Esq., M.A., M.D.　"Proceedings of Scottish Society of Antiquaries," 1883-84, pp. 216-225.　This writer states that the organic remains found in the Kjökkenmöddings are as follows :—

1. *Shellfish.*—Oyster, cockle, and mussel (most common), *Venus palustra, V. aurea, Trigonella plana, Nassa reticulata*, and *Littorina littorea* (most common of their kind), *Littorina obtusata, Buccinum undatum, Helix strigella, H. nemoralis*, and *Carocolla lapicida.*

2. *Fish.*—Herring, cod (*Gadus callarias* and *œglefinus*), eel, and flounder or dab.

3. *Birds.*—Eagle, cormorant, mew, wild duck and goose (most common), swan (*Cygnus olor et musicus*), capercailzie (*Tetrao urogallus*), and great auk (*Alca impennis*).

4. *Mammalia.*—Stag, roedeer, and wild boar (most common), urus (*Bos primigenius*), dog, fox, wolf, marten (*Mustela martes et foina*), hedgedog, otter, seal, porpoise, water-rat, mouse, beaver, wild cat, lynx, and bear (*Ursus arctos*).

5. *Vegetable Remains.*—Except ashes and charcoal, the latter of which on being analysed was found to belong mostly to a species of pine, and the charred remains of some kind of sea plant, no other products of the vegetable kingdom were found in any of the Kjökkenmöddings.

From the above list it will be observed that, except in the solitary instance of the dog, the ordinary domestic animals, as the common barn fowl, domestic ox, horse, sheep, goat, and domestic hog, are unrepresented.　In addition, we have also to note the absence of the mammoth and all the other extinct or emigrated mammalia of the Palæolithic period, including the reindeer, bison, moosedeer (*Cervus alces*), musk ox, and hare.

[16] Professor Steenstrup informs us in a letter, dated 15th March 1885, "that excavations were carried on at both Gudumlund and Sölager several times prior to 1873 without any remains of *Alca impennis* being found."

to prove conclusively that the Great Auk was more than a casual visitant of its shores.

About 150 of these Kjökkenmöddings are now known in Denmark.[17] Up to the year 1869 only about 40 of these had been examined[18] by the committee of investigation appointed by the Royal Society of Sciences of Copenhagen. This committee consisted of Professor Steenstrup, Dr. Worsäe, and the late M. Forchhammer, representing the respective branches of science, biology, archæology, and geology. Since 1869 a few more of those kitchen-middens have been examined, but it is evident there is much work yet to be done.

No one can doubt that we only stand on the threshold of this branch of inquiry as to Great Auk archæology, and it is to be hoped that the encouragements of the past may induce the younger archæologists of Denmark to follow in the footsteps of their worthy predecessors with an enthusiasm as great as that which has led to the attainment of such splendid results in the past.

ICELAND.

On the 21st of May 1858, Mr. (now Professor) A. Newton and Mr. J. Wolley arrived at Kyrkjuvogr, in Iceland. The following day the latter gentleman picked up from a heap of blown sand two or three humeri of the *Alca impennis*. This led to further search being made by both in all likely localities during their stay, but with variable and sometimes disappointing results, as, when their researches caused them to excavate some ancient kitchen-middens, where naturally they expected to discover bones of the Garefowl, they were invariably disappointed; and it was only by careful observation when travelling about that they found some of the bones they had been looking for. In the wall of the churchyard at Kyrkjuvogr several bones of the Garefowl were got sticking in the turf which is used to bind the walls together, and finding this turf had been cut from a small hillock close by, it was searched, with the result that among a large number of bones of other *Alcidæ* were discovered several of the *Alca impennis*, L.

The greatest find was, however, at Baejasker, where Mr. J. Wolley one day as he was riding along called out that he saw two bones of the Garefowl lying upon the ground. On dismounting he found them to be the distal ends of the *humeri*, and apparently a pair; going to the spot, Professor Newton found a

---

[17] " Early Iron Age," p. 2. Engelhardt.
[18] *Compte Rendu* " International Cong. d'Anthro. et d'Arch.," 4th session, p. 135.

radius of the same bird. This locality was carefully examined on two other occasions, and some other remains recovered, and it is believed that these must have belonged to at least eight individual birds.[19]   From what Professor Newton tells us, he and Mr. Wolley apparently got bones representing at least two Gare-fowls in the heap of blown sand at Kyrkjuvogr, also in the wall of the churchyard of the same place, and in a small hillock in the vicinity, a number of bones were found presumably representing several Garefowls, and then we are told that at Baejasker bones representing at least eight individuals of the same bird were obtained. What became of all these bones we are not told; but, in a valuable paper written in 1870, the learned Professor only enumerates bones representing eight individuals, and states they are in his own collection, and mentions they were got by Mr. Wolley and himself in Iceland, and quotes the *Ibis* for 1861, which we have just referred to for information regarding them.[20]

---

[9] Professor A. Newton's Paper on Mr. J. Wolley's "Researches," pp. 394–6, *Ibis*, vol. iii., 1861.

[20] Professor A. Newton on Existing Remains of the Garefowl, *Ibis*, April 1870, p. 260.

CHAPTER VI.

BRITISH REMAINS OF THE GAREFOWL.

WE next come to consider the remains found in Britain; and we have now three localities—two in Scotland and one in England—where unmistakable traces of the Garefowl have been met with. These are Keiss in Caithness;[1] Oronsay, one of the Southern Hebrides;[2] and Whitburn Lizards, in the county of Durham.[3] As far as is yet known, these are the only places in Britain where remains of the Garefowl have been found.

KEISS.

During the year 1864, in the course of some excavations carried on at Keiss at the expense and under the superintendence of Samuel Laing, Esq., M.P., a quantity of remains were discovered in an ancient kitchen-midden, and were handed to Mr. Carter Blake, then the assistant secretary of the Anthropological Society, London, who called in the assistance of Mr. Wm. Davies of the Natural History Department of the British Museum. This gentleman found that among the bones were those of birds which he could not name with satisfactory accuracy without comparing with recent skeletons, and with Mr. Blake's permission he took them to the British Museum, and identified their respective species;[4] but there were some bones that he could not identify as belonging to any northern bird of similar size. The idea then occurred to him that they were remains of the Great Auk (*Alca impennis*, L.); and a comparison with the bones of the razorbill (*Alca torda*, L.) confirmed this impression; but unfortunately at that time there were none of the larger bird in the museum. He mentioned his belief

---

[1] "Prehistoric Remains of Caithness." By Samuel Laing, Esq., M.P., and F.S.A. Scot. With Notes on the Human Remains by Thomas H. Huxley, Esq., F.R.S., Professor of Natural History Royal School of Mines. London, 1866.

[2] "Notice of the Discovery of the Remains of the Great Auk or Garefowl (*Alca impennis*, L.) on the Island of Oronsay, Argyllshire." "Linnean Society's Journal, Zoology," vol. xvi. pp. 479-487.

[3] "Natural History Transactions of Northumberland and Durham," vol. vii. part 2, 1880, pp. 361-364.

[4] "Journal of the Anthropological Society," vol. iii. p. 34.

to Mr. Gerrard of the Zoological Department, and that gentleman informed him that he had recently mounted some bones of the Garefowl, and that he thought they were in Professor Owen's room.   Mr. Davies then went to the Professor and asked him if he would kindly let him see them ; but they had been returned to their owner, Mr., now Professor A. Newton, and were the first remains he received from Funk Island, through the Bishop of Newfoundland.[5]   Mr. Davies then communicated to Professor Owen his belief as to the remains from Keiss, and showed the bones to him, mentioning to whom they belonged, and how they came into his hands, with the result that the Professor immediately requested to be allowed to take charge of the specimens, and also the other bird bones that Mr. Davies had named, and then entered into a correspondence with Mr. Laing.   The result was that the bones supposed to belong to the *Alca impennis* were positively and authoritatively identified, which could be done by the Professor, fresh from a recent study of the osteology of the bird.   They are fully described by Mr. Laing in the book he afterwards published in 1866, entitled, " Prehistoric Remains in Caithness," and are also mentioned in papers read by the late Dr. John Alexander Smith before the Antiquarian Society of Scotland at their meetings of January 1867, January 1879, and June 1880.

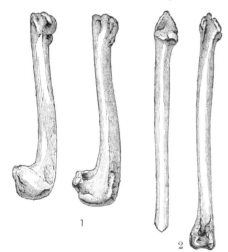

The bones consist of a right and left humerus, each measuring four inches in length, and perfect.   There is also a right and left tibia ; but only one of these is whole, and measures five inches in length, the other being without the lower articulation.   In addition to these there is another left tibia, but belonging apparently to a smaller bird.   This bone wants its lower extremity, and Professor Owen thinks it may have belonged to a female ; while the other bones are those of a male, and a part of the anterior or free extremity of the premaxillary of a Garefowl.[6]

1. Two Humeri ; and, 2, two Tibiæ of the Great Auk (*Alca impennis*), found in a kitchen midden at Keiss, Caithness-shire (half the natural size).

---

[5] " Annals of Natural History," third series, part 14. " Proceedings of the Zoological Society," November 10, 1863.

[6] These remains are now preserved in the Scottish National Museum of Antiquities, Edinburgh.

About the same time and place Mr. (now Dr.) Joseph Anderson of the Scottish National Museum of Antiquities, Edinburgh, found part of the sternum of one of these birds, which he gave to Mr. G. W. Busk, at that gentleman's request, as a donation to the Museum of the College of Surgeons, London; but it would appear that it is entered in their books (1150 B) as being a donation from Mr. Busk, and not Dr. Anderson, and it is but right that this mistake should be rectified.[7]

A most remarkable printer's error occurred in describing the upper mandible or premaxillary in Mr. Laing's book, and was repeated in the "Transactions of the Scottish Antiquarian Society," 1867, where it is mentioned among the remains obtained from Keiss, and figured as being a curved spear-head. It was only when the late Dr. J. Alexander Smith came to prepare his second paper, read before the same

Upper Mandible of Great Auk found at Keiss (half the natural size).

Society during 1879, that in examining the collection from Caithness he observed his mistake, and when his communication was made drew attention to it.[8]

As it is probable we are only on the threshold of investigation as far as ancient kitchen-middens are concerned in Britain, it will be interesting to observe the fauna of the Keiss deposit as given by Mr. Laing :—

MOLLUSCA.
  Limpet (*Patela vulgaris*).
  Periwinkle (*Littorina nontridia*).
  Whelk (*Buccinum undatum*).
  Cockle (*Cardium*).
  Scallop (*Pecten majus*).
  Lesser scallop (*Pecten argus*).

ANNULOSA.
  Lobster (*Serpula*).

FISH.
  Cod (*Morrhua vulgata*).

MAMMALIA.
  Ox (*Bos longifrons*).
  Horse (*Equus caballus* (?) *fossilis*).
  Red-deer (*Cervus elephus*).
  Goat (*Capra hircus*).

MAMMALIA (*continued.*)
  Hog (*Sus scrofa*).
  Dog (*Canis familiaris* or *familiaris fossilis*).
  Fox (*Canis vulpes*).
  Rabbit (*Lepus Cuniculus*), perhaps recent.

CETACEA.
  Grampus (*Delphinus orca*) or small whale.
  Dolphin (*Delphinus delphis*) or some other small cetacean.

BIRDS.
  Great Auk (*Alca impennis*).
  Lesser Auk or razorbill (*Alca torda*).
  Cormorant (*Phalacrocorax carbo*).
  Shag (*Phalacrocorax graculus*).
  Solan Goose (*Sula bassana*).

---

[7] "Proceedings of the Antiquaries of Scotland," vol. i., new series, 1879, p. 81.   [8] *Ibid.* p. 79,

Dr. J. Alex. Smith has also identified a portion of the antler of a reindeer.[9]

In addition to these remains, Mr. (now Dr.) J. Anderson recovered bones representing the following, which were named by Mr. G. W. Busk :—

BIRDS.

The large Guillemot. | The Great Auk (*Alca impennis*).

MAMMALIA.

A small Bos Taurus. | Small mature sheep.
A very large fox. | A very young lamb.

All these remains are thought to be comparatively recent, and to belong to the late period of the occupation of the Broch near which they were found; but as there is a bone of the reindeer, which probably became extinct in Scotland about seven centuries ago, as it is last mentioned in the "Orkneyinga Saga" in the year 1159 A.D., the deposits most likely had been formed at a period previous to that time.

---

[9] "Proceedings of the Society of Antiquaries of Scotland," vol. i. new series, 1879, p. 76, 77.

CHAPTER VII.

BRITISH REMAINS OF THE GAREFOWL—continued.

ORONSAY.

THE most recent discovery of bones of the Garefowl must now come under consideration, and we shall endeavour to give an account of the events that led
up to it.

During the summer of 1879 the writer formed one of a small party who
visited the Island of Colonsay, to which is attached at low water the Island of
Oronsay, the intervening strand being dry for about three hours each tide. The
islands were so interesting that we felt we had entered upon a new field for study;
accordingly we began to make a list of the flora (since published by the Botanical
Society of Edinburgh),[1] and in the course of our rambles we endeavoured to
note all the most interesting features of the surroundings, whether natural or
antiquarian.

At the beginning of May 1880 we returned to the islands a second time,
and were struck with the remarkable appearance of a cone-shaped mound on the
eastern side of Oronsay. We shortly afterwards discovered that Pennant, when he
visited the island in 1772,[2] had noticed this place, and in his " Tour through the
Western Isles" described it as a tumulus. We at once resolved that if an opportunity afforded, and we could obtain the permission of the proprietor, we should
endeavour to find out by excavating what the mound, known to the islanders by
the name of Caisteal-nan-Gillean,[3] concealed. We did our best by inquiries
among the inhabitants to find out all that was related about it, either from tradition or actual knowledge, but the result of all our questioning was absolutely *nil*.

During the following winter we made the acquaintance of Mr. William
Galloway, known for his antiquarian researches and work in connection with the

---

[1] "Transactions of Botanical Society, Edinburgh," vol. xiv., part i. pp. 66-73; part ii. pp. 219-224.
[2] Pennant's "Tour through the Western Isles," 1772. Published at London. Vol. iii. p. 290.
[3] In the Gaelic, literally the castle of the servants (gillies or followers).

ancient ruins and sculptured stones of Scotland, and discovered that he had some years previously visited Oronsay when taking casts and drawings of many of the carved stones and crosses at the ruined priory. It was soon arranged that we should revisit the island, and at the beginning of June 1881 Mr. Galloway started, and was followed about the middle of that month by the writer. On going on board the steamer *Dunara Castle* at Greenock, we met Mr. Malcolm M'Neill, brother of Major-General Sir John Carstairs M'Neill, V.C., the proprietor of the islands, and had a conversation with him regarding Caisteal-nan-Gillean, and found that, acting for his brother in his absence, he was willing to give permission to open the mound. We visited it soon after arriving on Colonsay, accompanied by Mr. Alexander Galletly, curator of the Museum of Science and Art, Edinburgh, and were joined at the mound by Mr. Galloway, who was living at Oronsay.

Satisfied from our inspection that the mound was worth examining more carefully, we made some preliminary preparations and reported to Mr. Malcolm M'Neill, who then gave us his final sanction to begin excavating. We hired a workman, and on Wednesday the 22d June walked from Scalasaig on Colonsay to the mound on Oronsay, a distance of fully five miles. When we arrived at our destination with our shovels, we found Mr. Galloway waiting, as he was interested in our proposed work.

We began by making a survey of the whole mound, which led us to decide to commence operations on the eastern side, at a point where the wind had blown away part of the sand that appeared to form the greater part of the hillock, and cut an opening about four feet wide right through to under the apex. We commenced digging, and at first little was found to reward our efforts, but as we gradually worked inwards we came upon a thin layer of shells on the upper surface below the turf that covered the mound ; and as this gradually got thicker and we found intermixed with the shells a few bones, it raised our hopes, and it was then agreed that Mr. Galloway should join in the operations. It was determined that while the workman and myself carried on the excavation, Mr. Galloway was to measure and lay off a section of the mound with the ground to the south-east, where there is a sandpit, from which we supposed the sand which formed all but the outer crust of this remarkable hillock had been taken.

Our cutting was commenced at the base of the mound, and as we dug towards the centre of it we slowly formed a deep trench. We found the work rather dangerous, as large quantities of sand were constantly falling, and the walls of the trench rose considerably above the level of our heads on either side.

At last we found it impossible to work straight in upon the same level on which we started, and had gradually to work up an inclined plane, so as to keep the bottom of our trench about ten feet from the surface of the mound as we steadily excavated towards its middle. We found the work rather heavy, and to add to our difficulties the man we had engaged, though a stout Highlander, declined to come back after one day's trial. However, the following morning we succeeded in engaging another workman, with whom we had done some of the excavating at the Crystal Spring Cavern, and we found things go on much more satisfactorily as his previous experience in looking for remains was of some value.[4]

While we were engaged digging, Mr. Galloway was busy measuring and marking off the ground. He ascertained that the hillock was 150 feet in diameter, and nearly circular in form; the height being about 30 feet on the eastern side, which gives the greatest elevation, and about $21\frac{1}{2}$ feet on the western side, as the ground rises considerably in that direction, the mound having been formed on links that slope towards the sea.

View of Caisteal-nan-Gillean.

At the end of three days we had made a cutting or trench about 70 feet in length, and were close to the apex, with the result that we had discovered abundance of shells, a few bones, and some rough stone-implements; and with these we started for Edinburgh. After carefully examining all the material we had collected, we picked out those bones that were least fragmentary, and having divided them into two lots, we placed the first of these in the hands of our friend Mr. Alexander Galletly, Curator of the Museum of Science and Art, who was kind enough to enlist the aid of Dr. R. H. Traquair, F.R.S., Curator of the Natural History Department of the same Museum, in their identification. In the meantime we obtained the promise of help from Mr. John Gibson, assistant to Dr. Traquair, with the second lot, which, as it afterwards turned out, contained the remains about which we now write.

On receiving the remains Mr. Gibson at once began their identification, and in the course of the work showed the bones to Dr. Traquair, who was also interested. It would appear that both these gentlemen had been impressed with the remarkable form of one of the bird humeri, but made no observation to each other. This was on a Saturday. On the Monday morning following they met in the Museum, and proceeded in the direction of the case in which were preserved some remains of the Great Auk that were obtained at Funk Island by Professor J. Milne. In conversation they discovered that they had both got the impression that this humerus belonged to a Great Auk, and that without any previous comparison of opinions they were both proceeding to have a look at the remains in the Museum, so that they might, if possible, verify their surmise. Fortunately they found that they were correct in the opinion they had simultaneously formed, and the identification of the one bone led to the discovery that several of the other bones also belonged to the Great Auk.

This discovery gave so much encouragement that it led us to make arrangements to return to Oronsay in August of the same year, and continue the excavations. Mr. Galloway started about the middle of the month, and remained working for six or seven weeks, aided by two boys, whom he succeeded in engaging. We were fortunate in being able to spend about a week in his company, and though every effort was made to secure the services of one or more men, we were unsuccessful, as the harvest fully occupied those who remained at home, and many of the islanders were at the fishing.

During this visit we were employed removing the upper part of the mound, where the greatest deposits existed; as our experience showed us that if it had been raised over anything, or was the superstructure covering a place of interment, we could only ascertain this by digging down to the living rock, which is about three feet below the original level of the sand at the outer edge of the mound, and possibly is the same under the apex. As the sand falling would make this work very dangerous, if not impossible, we resolved first to remove about twelve feet off the upper part of the hillock, and then dig downwards, as circumstances permitted.

By the end of September fully one-third of the apex had been dug off, and every spadeful most carefully examined, so that not even minute objects could be passed. The same care was taken during the whole time the excavations were in progress, and though the work went on much more rapidly during our first visit, it was entirely owing to our having mostly pure sand to deal with, which contained not a vestige of remains, and seemed as if just deposited from the sandpit.

To give some idea of the nature of the deposits as revealed by the sections examined during the digging, we may state that the outside of the cone is covered with turf and blown sand to a depth varying from one to five feet, the greatest depth being at the north side of the apex,[5] and gradually thinning off all round to the outer edge; below that is a series of strata, composed principally of shells, which taper off from the apex similarly to the upper deposits, and underneath is pure sand.

Where we began our excavations we found almost solid sand; then after a few feet we came upon a thin layer of shells near the surface, which was at first only about an inch thick. As we worked inwards this line was found gradually getting thicker, until near the summit it was composed of numerous layers, which were pretty clearly defined, though here and there they ran into each other, and altogether were about 8 feet from top to bottom.

The greater part of the shells were those of the Limpet (*Patella vulgata*, L.), others were, however, intermixed, of which we give a list amongst the remains (see p. 55). Besides these there were a few bones, bone implements, and oblong water-worn stones of a slaty character, some of which we suppose have been used as limpet hammers, while others have one end rubbed so as to form an edge, and are similar in appearance to implements of a like kind we have seen from the Swiss lake-dwellings, and also from a number of places in our own country. There are also a few oval and nearly round stones that have marks that lead one to suppose they have been used for rubbing; and others seem to have been implements used for striking the head of a chisel or other similar tool, as there are well-defined

---

[5] The strong winds from the south and south-west that blow over the island have caused the accumulation on the north side of the mound.

indentures that indicate the point of contact. We also got some stone-heaters that have been cracked by the action of fire, and in addition a few pieces of flint of small size. Of bone implements we got several, but all in a fragmentary state. They consisted of a number of barbed harpoon or spear-heads, one bone awl in a perfect state, and the point of another; also a number of bones rubbed at one end, some on both sides, so as to form an edge, and others only on one side; but most likely they were used for different purposes, as those rubbed flat only on one side are larger, and made of selected pieces of the bones of red-deer, while some of those with the rubbing on both sides, so as to form an edge, are made of the same material, but portions of smaller bones have been used. In digging we discovered some large flat stones which had evidently been used as hearths, for they had charcoal and burnt material around them, but not in sufficient quantity to give the impression that they had been used for any great length of time, and it was generally in the immediate neighbourhood of those ancient fire-places that we got the implements. The charcoal is very soft, and has the appearance of having resulted from the burning of a soft wood. In the bed of Loch Fada on Colonsay are stumps of immense trees that may at one time have furnished the inhabitants with fuel.

As we were anxious to find out whether the charcoal and the wood from these tree-stumps agreed in structure, we placed specimens of each in the hands of Dr. J. M. Macfarlane, assistant to the Professor of Botany in the University of Edinburgh. He informs us that though the charcoal is in such a state that it is beyond identification, he has been able with some certainty to identify the wood as having belonged to the goat-willow (*Salix caprea*, L.); he also thinks that the charcoal is the result of this wood being burnt, as they have some characteristics in which they correspond. It is probable Dr. Macfarlane is correct, as the willow was much used in the Hebrides for making bridles, ropes, and tackle of every variety.

The remains in the lowest deposits near the summit differed in some respects from those found nearer the surface. All are of a very rough description, indicating that this mound was used by a primitive and probably ancient people. In fact, the question naturally arose, What could there be underneath that would account for the sand-hill?

Later excavations, carried on during the month of March 1882, by Mr. Galloway, have shown that the sand below the strata in which we had found the remains is not one vast homogeneous mass that has been accumulated at one time, but is all blown or drift-sand, laid in regular layers, the upper part of each defined by a thin line of dark mould, with a few sea and land shells intermixed, but no implements or other remains have yet been met with in the lower deposits. The conclusion that all seems to point to is, that the lower part of the Caisteal-Nan-Gillean has been formed by natural and not human agency.

The following is a list of the remains in our possession that have been already identified. We are indebted to Dr. Traquair for identifying the remains of the fish, and also those of the seal and pig; while to Mr. John Gibson we owe the determining of the others, and the description of the bones of the Garefowl :—

Remains.

Bones of the Great Auk or Garefowl (*Alca impennis*, L.) obtained from Caisteal-nan-Gillean, Oronsay, June and September 1881.

I. *Right humerus*, measuring 4 inches in length and 1 inch in breadth at the proximal end. The compressed shaft at the middle portion measures 6 lines in long diameter and 3 lines in short diameter.

According to Professor Owen ("Trans. of Zool. Society," vol. v. p. 327), there is a thick ridge or raised rough surface near the radial end of the articular head of the humerus, extending about 8 lines down the bone, which gives insertion by a well-marked narrow elliptical depression to the second pectoral muscle—the raiser of the wing.

In the present specimen the bone of this ridge exhibits a diseased condition, the normal depression being changed into a deep trough 8 lines in length and 4 lines broad.

II. *Proximal half of right humerus.* Total length of specimen, $2\frac{1}{2}$ inches, broken about the middle of the shaft, which exhibits medullary cavity. This cavity measures $2\frac{1}{2}$ lines in long diameter by 1 line in short diameter, the shaft measuring similarly 6 lines by $2\frac{1}{2}$ lines.

III. *Distal half of left humerus.* Specimen, measuring 2 inches 2 lines, shows medullary cavity. In this specimen the condyle and the three anconeal ridges are very perfect.

IV. *Distal end of left humerus*, 3 inches in length.

V. *Left coracoid bone*, with a total length of 2 inches 4 lines. At the sternal end it is 10 lines in breadth; but as both ends are imperfect, it probably had a breadth of at least 1 inch. The thin lamelliform process given off above the sternal articulation is also gone, otherwise the coracoid is entire. From the sternal end it gradually contracts to 5 lines, then widens out, giving off a strong compressed process, which is perforated.

VI. *Upper end of right coracoid.* Specimen 1½ inches in length, ending a little below the perforated process.

VII. *Distal end of right tibia.* Specimen 1 inch in length. Shaft showing very minute medullary cavity.

VIII. *Dorsal vertebræ.*

Other Remains.

MAMMALIA.

Red-deer (*Cervus elaphus*, L.); many of the fragments have been rubbed, and all the bones have been broken.

Marten (*Martes foina*, L.)

Otter (*Lutra vulgaris*, Erxcl.)

Sheep (*Ovis aries*, L.); we have only one portion of a bone that we are certain belongs to this animal, and it was found near the upper surface of the deposit under the turf. It is in better preservation than the other remains, which may indicate that it is more recent.

Rat (*Mus decumanus*, Pall, or *rattus*, L.)

Rabbit (*Lepus cuniculus*, L.), found in old burrows. The remains appear to be recent.

Common seal (*Phoca vitulina*, L.)

Wild boar or pig (*Sus scrofa.*)

BIRDS.

Guillemot (*Uria troile*, L., or *Grylle*, L.)

Razorbill (*Alca torda*, L.)

FISH.

Wrasse (*Labrus maculatus*, Bl.)

Grey mullet (*Mugil septentrionalis*, Gunth.)

Picked dog-fish (*Acanthias vulgaris*, Risso).

Skate (*Raja batis*, L.)

CRUSTACEANS.

Crab (*Platycarcinus pagurus*, Edw.)

SHELLS.

Limpet	(*Patella vulgata*, L.)			(*Axinœa glycymeris*, L.)
Scallop	(*Pecten opercularis*, L.)		Cockle	(*Cardium edule*, L.)
Oyster	(*Ostrea edulis*, L.)			(*Tapes pullastra*, Mont.)
Horse whelk	(*Buccinum undatum*, L.)			(*Tapes virgineus*, L.)
Periwinkle	(*Littorina littorea*, L.)			(*Venus casina*, L.)
	(*Cyprina islandica*, L.)			(*Ensis siliqua*, Linn.)
	(*Lœvicardium norvegicum*, Spengl.)			(*Trivia europœa.*)

Besides the above remains which came into our hands, Mr. Galloway exhibited the following at the International Fisheries Exhibition, held in London during the summer and autumn of 1883 :—

MAMMALIA.

Wild boar or pig (*Sus scrofa*), two tusks which have probably belonged to that animal.

Otter (*Lutra vulgaris*, Erxl.) four fractured jaws and seventeen other bones.

Grey seal (*Phoca gryphus*), four bones.

Common seal (*Phoca vitulina*, L.), nine bones ; also three teeth, one of which may belong to the grey seal.

Red-deer (*Cervus elaphus*, L.), five or six fragments of bones ; each bone has been hacked all round and broken across.

CETACEANS.

There are three cetacean bones that have probably belonged to the *rorqual* or *finwhale.* They are : One large fragment of a rib, one part of vertebral epiphysis, and one fragment of a rib made into a pointed spear or lance-head.

BIRDS.

Great Auk (*Alca impennis*, L.). The proximal halves of three right humeri. The proximal half of one left humerus. One perfect coracoid bone, and one fragment of another coracoid bone. Two fragments of tibias, a proximal and a distal end ; one imperfect bone, said also to be from this bird.

Wild swan ; there were several bones said to have belonged to this bird.

In addition to the foregoing there was exhibited a collection of fish vertebrae, some crabs' claws, and a number of shells, representing some of the varieties already given in our list.

A number of implements were also shown : they were eleven barbed bone spear-heads; a large number of rubbed stones and bones; also a collection of limpet-hammers, a few flint chips,, some large flat stones, supposed to be lap-stones, as they are said to bear indentations, made, it is thought, by bones being split on them. It appears to us that the three largest are the most interesting. Two of these stones are oblong and one circular, the latter being artificially chipped into its present form.

With regard to the whole of these remains, we may observe that the bones of the red-deer, though found all through the strata, even in the highest, were most plentiful in the lower deposits, and seemed to become much less common in the upper layers. This probably indicates that the animal was becoming gradually less abundant during the period that Caisteal-nan-Gillean was inhabited. We may also remark that, as in our excavations at the Crystal-Spring Cavern, Colonsay, we only found the bones of the red-deer in the lowest deposits of the cave-floor. It appears, therefore, that there is good reason to suppose that the time at which the upper deposits at the mound and the earliest deposits of the cave were formed, is about the same. Moreover, as the deer-remains in the cave are not found very frequent, it is quite possible it was only occupied after the mound had ceased to be a place of human residence. If our conclusions upon this point be correct, the mound must have been occupied at a very early period in the history of the isles, as we find in the upper deposits of the cave-floor, and above the strata in which we have found the deer bones, other remains which point to these deposits having been formed during the Danish and Norwegian occupation of Colonsay and Oronsay.

Another evidence of the antiquity of the mound is in the absence of ox-remains, which are met with under stalagmite, in the upper deposits of the cave. The remains described as those of the pig may possibly belong, not to the domestic hog, but to the wild boar. As Sir John Lubbock remarks,[6] " Professor Steenstrup does not believe that the domestic hog is represented by its remains in the Danish shell-mounds." Besides, one of the rib bones in our possession bears evidence of having been broken and afterwards having united, and such an injury, it seems to us, would most likely be received in the chase.

One remarkable feature of the deposits at Caisteal-nan-Gillean is the immense number of limpet-shells, very many with small holes in them, caused, we believe, by the stroke of the rough stones used as hammers to knock them off the rocks.

---

[6] "Natural History Review," 1861, p.497.

Almost all the stone-implements are just suitably shaped stones taken from the beach; but nearly all those found in the neighbourhood of the hearths bear marks of having been rubbed at the one end, and, with two exceptions, are all small, varying from two to three inches in length, while many of the stones we call limpet-hammers are quite a foot in length, and with the exception of being sometimes fractured at the ends, bear no evidence of having been used. Nearly all these were found lying among the thinner deposits of shells away from the centre of the mound, as if they had been thrown there to be out of the way from the hearths.

Our reason for calling them limpet-hammers is as follows:—We had been making inquiries among the islanders for those implements, but without success, as we understood they were carefully fashioned or selected stones that were handed down by the fishermen from father to son, and we found that most of the men used the blade of an old reaping-hook to knock the limpets off the rocks. However, we also discovered that failing an instrument of that kind, they then took an oblong-shaped stone from the beach. The second day of our excavations at Caisteal-nan-Gillean we were puzzling ourselves as to what could be the use of the numerous oblong stones we met with among the shells, and mentioned the matter to our workman, who was accustomed to go to the fishing, and he, just as a matter of course, informed us they were limpet-hammers. He assured us that he and his fisher-mates often took such stones from the beach when proceeding on a trip, and would retain the stone for collecting bait until the end of their fishing, when they would throw it away. Subsequent inquiries have only helped to confirm us in the opinion that the large oblong stones found at Caisteal-nan-Gillean are really limpet-hammers. We understand that similar stones have been found in the ancient kitchen-middens of other localities, and have proved a puzzle to antiquarians; but we think what we have stated will be found to be the real solution of the mystery.

The bones of the Garefowls were found intermixed with other remains in this kitchen-midden of the ancient inhabitants, and this leads us to the conviction that the birds had been used as food. It is exceedingly probable that the Garefowls bred upon the rocky islets that lie near the shores of Oronsay, if not upon that island itself, and would prove an easy prey to its inhabitants. Though the bones of this bird as yet found here are few in number, it does not necessarily follow that the Garefowl was not very plentiful at this station, as not one bone in a

H

thousand would be deposited under such conditions as would enable it to resist the ravages of time. In fact, the evidence is all in the other direction. We might reasonably expect to find the Alcine crania and many other bones where we found the humeri; but we have not found any trace of such remains, and it is only the thickest and strongest bones that we have discovered, and these bear evidence of great age.

CHAPTER VIII.

HOW WAS CAISTEAL-NAN-GILLEAN FORMED, AND TO WHAT PERIOD DOES IT BELONG?

THE excavations at Caisteal-nan-Gillean have revealed that it is a place on which some of the ancient fisher-folk of Oronsay dwelt, and they probably chose it as a place of residence because possessing two great advantages—namely, a dry soil, and an extensive prospect seaward. The former of these was a most important consideration, as the inhabitants evidently did not live in stone buildings, and if they had any houses to cover them, these were most likely made of wattles. The extensive view of the sea was imperative for a fisher population, and if it enabled those at home to see the returning boats in the far distance, and provide in good time the morning or evening meal, it also afforded a vantage-ground from which the approaching enemy might be seen in time of tribal strifes and petty wars. From it also might be observed the signs upon the surface of the sea that betokened the shoal of fish, or a view could be obtained of the sea-birds resting on the rocky shores of Eilean-Ghurdimeal and Eilean-nan-Eon, or floating on the troubled waters of the wild Atlantic; in short, from Caisteal-nan-Gillean its inhabitants could watch for friend or foe, for food or storm.

It has been said that Caisteal-nan-Gillean is part of an ancient raised sea-beach, and the deposits upon it are supposed to have been formed when it was near the sea level. As appearing to confirm this supposition, it has been advanced that in adjoining sand-hills and sand-wreaths, where recent storms have not changed the old contour of the sand dunes, layers of sea-shells have been found which correspond with layers found at similar elevations upon them all, and it has thus been concluded that these layers could only have been so deposited either by the sea or by human agency. From such conclusions we beg respectfully to differ, and as the settlement of those questions have an important bearing on the period of time to which we must refer the formation of the deposits at Caisteal-nan-Gillean, we shall briefly state our views and the reasons which have led us to adopt them.

During the excavations we had occasion to dig down to the schistose rock underneath the sand at the base of the mound, and we found it, not rubbed to a smooth surface as it would have been if exposed to the wash of the sea, but with a rough surface, bearing evidence of having been exposed as dry land to the influence of the weather for a considerable period before it was covered with blown sand. As we discovered no pieces of stone that had been broken from the rock strata by the effects of weather, and not the slightest trace of the decomposition of such fragments beneath or among the sand, we think this is evidence that the rock had been exposed as dry land to the action of the weather for some time before it was covered up. If this is so, it follows that Caisteal-nan-Gillean and the adjoining sandhills are composed entirely of blown sand, and even if there were nothing else to judge from, their contour might almost lead to this conclusion. Any sea-shells that are to be found have doubtless been blown into their present position by the wind, which comes with tremendous force off the Atlantic during the frequent gales that occur in these Western Isles. If any of our readers have stood on the sandy shores or links of one of the Hebrides in time of storm, and seen the blinding drift of sand and shells blown along by the gale, they may have observed, that as the storm gradually abates the wind is unable to carry along with it the heavier objects, and that the sand is blown away from the shells, which are displayed in a layer upon the surface of the beach; or if they have visited the shores of Holland, where sand-dunes fringe the coast, and keep back the encroachment of the sea, they may have seen the same operation performed by the wind under similar circumstances. If such has been their good fortune, we think they will have no difficulty in accounting for the scattered shells on the upper surface of the layers of sand under the human deposits on Caisteal-nan-Gillean, and also for those on the sandhills and sand-wreaths adjoining.

We admit that at one time the part of Oronsay where Caisteal-nan-Gillean stands was under the sea, and it was during the process of elevation of the land, probably a gradual operation, that all the rocks must have been exposed to the wash of the waves. As the land rose, the rocky shores which gradually rise inland from the present high-water mark were eroded by the effects of the weather, as evidenced by the rock at the base of the mound. We therefore think our readers will agree with us that there seems good proof that it was at a time long after this rock had ceased to be washed by the waves that it was covered with blown sand, which was formed at this part of the island into sandhills and wreaths with the eddying winds.

If our supposition is correct, it will be seen that it is not necessary to relegate the objects found on Caisteal-nan-Gillean to a period much earlier than the Christian era, as would be requisite if we suppose the remains to have been deposited when the part of Oronsay on which Caisteal-nan-Gillean is situated was at or near the sea level, and formed part of the sea shore. Moreover, the opinion we have stated as to the probable period of the formation of the mound seems to be borne out by the dim light of early Scottish history.

During part of the past summer (1884), Mr. W. Galloway has been excavating at a second shell-mound on the island of Oronsay, and among other remains has discovered a coracoid bone of *Alca' impennis*. From the situation in which this bone has been found, we conclude that the bird to which it belonged was used as food. We think it is also additional evidence that at one time the Garefowl was common in the neighbourhood of Oronsay, and probably bred upon the numerous rocky islets near its shores.

CHAPTER IX.

ENGLISH REMAINS OF THE GREAT AUK.

THE discovery of traces of the Great Auk in a cave near Whitburn Lizards, county Durham, during the spring of 1878, is very interesting, as until that time no remains of this bird, so far as known, had been found in England. There can be little doubt that at one time the Great Auk was in the habit of visiting the shores of even the most southern parts of Britain, but it is long since these visits became of very rare occurrence. The last notice that we know of the Great Auk having been met with in the north-east of England, is the mention that a specimen had been captured on the Farn Islands about a century ago.[1]

It appears that the workmen employed by the Whitburn Coal Company had been quarrying limestone on the eastern escarpment of the Cleadon Hills, named on the Ordnance Survey Map " Whitburn Lizards," when underneath a quantity of *debris* which had at one time fallen from the face of the cliff, they discovered a cave which at some remote period had evidently been formed by the sea when the land was at a lower level, as it was situated on the north-east escarp of the hill, about fifteen feet from its summit, and 140 feet above the present sea-level. Mr. Howse, who was one of those who examined it, has written a preliminary description of the cave and its contents.[2] He states that he believes this cave, along with other two adjoining it that have since been discovered, were raised to their present elevation long before being occupied by the creatures whose remains have been found in them, and that probably the deposits on the cave-floors are not of extreme antiquity, as in none of them were discovered traces of the hyæna and cave-bear, met with in such abundance in some other English caves. The cave ran nearly due west into the hill, bifurcating near its inner end. The entrance was about four feet high before the strata of the cave-floor was disturbed.

---

[1] " Catalogue of the Birds of Northumberland and Durham," by Mr. John Hancock. " Transactions, Tyneside Naturalists' Club," vol. vi. p. 165.

[2] " Natural History Transactions of Northumberland, Durham, and Newcastle-on-Tyne," vol. vii. part ii., 1880, pp. 361-364.

The deposits consisted of a layer of cave earth, with an irregular surface that was grooved with the drainage of water, and was of reddish colour, rather fatty to appearance, but friable when dried. It averaged about two feet in thickness, was intermixed with the various remains, and rested on the soft, nearly yellow lime-stone that formed the bottom of the cave. It seems rather strange that no rounded water-worn stones were found, as these are generally present on the bottom of sea-caves; and it is by the grinding power of these implements, when set in motion by the waves, that caves are usually formed along the lines of veins of softer rock (frequently limestone) that run through cliffs; but in the case under notice they are conspicuous by their absence.

Until this discovery the scientists acquainted with the locality had no idea of the existence of any caves in the neighbourhood, and it must have caused considerable surprise to the officials of the Museum of the Natural History Society, Newcastle-on-Tyne, when, in the spring of 1878, they received the first box containing the remains, which were kindly sent them by Mr. John Daglish, Tynemouth, who at the same time gave liberty for some members of the Society to excavate in the cave. It was fortunate that such a competent authority as Mr. John Hancock undertook the examination of the remains, as his labours have resulted in the identification of bones that have belonged to a considerable number of mammalia and birds, along with the shells of several of the mollusca.

Among the former of these it is worthy of notice that there are several domestic animals, but their remains are associated with those of some animals that have long been extinct in the North of England. It would be interesting to know if the first excavations, and those that were afterwards undertaken, were conducted with the scientific accuracy necessary to preserve the sequence of the various layers of cave earth, the age of which would of course be in proportion to their depth. We would therefore expect that the remains of the domestic animals might probably be mostly found among the upper layers of cave earth, or possibly upon its surface; while the remains of the wild animals now extinct in the locality would be most plentiful in the lower strata, and gradually decrease in quantity towards the upper layers, and possibly be entirely absent from the surface. However, as the paper that has been written by Mr. Howse is only a preliminary communication, there can be little doubt that when he gives a full account of the explorations he will refer to the subject.

The following is a list of the species whose bones or shells have been identified :—

MAMMALIA.

Horse, cow, sheep, dog, pig or wild boar, red-deer, roe-deer, badger, fox, yellow-breasted martin, weasel, hedgehog, mole, water-vole.

BIRDS.

Kestrel or merlin, gannet, great auk, razorbill.

MOLLUSCA.

Oyster, periwinkle, limpet, and several species of snails.

Of the above remains by far the most valuable and interesting is the single bone of the Great Auk, which has thus left a trace of its presence on the shores of this part of Britain in early times. It is fortunate that the bone, which is an upper mandible, is so characteristic that it cannot be mistaken for that of any other bird; and we are greatly obliged to Mr. John Hancock for kindly furnishing us with the drawing from which the accompanying woodcut has been prepared.

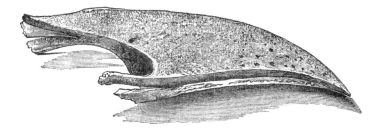

Upper Mandible of a Great Auk found in a cave near Whitburn Lizards, County Durham, during 1878 (natural size), from a drawing by John Hancock, Esq.

It is almost a wonder that amid such a quantity of remains this bone was so easily identified, and it is also evidence of how thorough is the disappearance

CHAPTER X.

THE HABITS OF THE GAREFOWL, AND THE REGION IT LIVED IN.

THE habits of the Garefowl appear to have led it to frequent those isolated situations where, under ordinary circumstances, it would be free from molestation by men, as the bird's want of the power of flight made it so helpless when on land. It is unfortunate that, owing perhaps to this instinctive retirement from places of human existence, we know really so little regarding it. One of the best descriptions we have is that by Martin, who, writing of St. Kilda,[1] says : " The sea-fowl are first the Gairfowl, being the stateliest, as well as the largest sort, and above the size of a solan goose ; of a black colour, red about the eyes, a large white spot under each, a long broad bill. It stands stately, its whole body erected ; its wings short, flies not at all ; lays its eggs upon the bare rock, which if taken away, she lays no more for that year. She is whole footed [web-footed], and has the hatching-spot upon her breast, *i.e.* a bare spot from which the feathers have fallen off with the heat in hatching,—its egg is twice as big as that of a solan goose, and is variously spotted, black, green, and dark. It comes without regard to any wind, appears the first of May, and goes away about the middle of June."

A later writer, who visited St. Kilda in June 1758, says : " The bird visits the island in July ; " but this is evidently a mistake, and he acknowledges he had not seen it himself.[2] Dr. John Alexander Smith refers to this, and seems to be of the opinion that in 1758 the Garefowl was breeding on some isolated skerry among the Hebrides, and was only seen at St. Kilda after the period of incubation was over.[3] But we think this seems hardly probable, as when we have any mention of the Garefowl being seen at St. Kilda, it is always at its breeding-time. What came over the bird at other periods of the year is likely to remain a mystery ; but though there was a generally accepted idea among the sailors and fishermen of

---

[1] "A Voyage to St. Kilda, May 29, 1697," by M. Martin, gent. Published, London, 1753, p. 27.

[2] "History of St. Kilda," by the Rev. Mr. Kenneth Macaulay, minister of Ardnamurchan, missionary to the island from the Society for Propagating Christian Knowledge. London 1764, p. 156.

[3] "Proceedings of the Scottish Society of Antiquaries," vol. i N.S., p. 90, 1878-79.

I

Newfoundland [4] that it did not leave soundings, it seems evident that it must have occasionally made long sea journeys. The stray birds that appeared from time to time at various parts of our coasts in this century, indicate that, as it did not breed during this period at any nearer point than Iceland, it must at least have found its way across the sea from there. If this was the case when the bird had become so rare, we may infer that it migrated in much larger numbers in early times when it existed in such strong colonies. It is said "that they swam with their heads much lifted up, but their necks drawn in ; they never tried to flap along the water, but dived as soon as alarmed. On the rocks they sat more upright than either the guillemots or razorbills, and their station was further removed from the sea. They were easily frightened by noise, but not by what they saw. They sometimes uttered a few low croaks. They have never been known to defend their eggs, but would bite fiercely if they had the chance when caught. They walk or run with little short steps, and go straight like a man. One has been known to drop down some two fathoms off the rock into the water. Finally, I may add that the colour of the inside of their mouths is said to have been yellow, as in the allied species." [5] Eggert Olafsson, writing in 1772, [6] says: "The eggs lie together on the bird's dung ; they build no nest. Several Garefowls have this nest and eggs in common." Professor Steenstrup, writing 13th April 1885, informs us that "Mr. E. Olafsson never saw the Garefowl living, or at its nesting place." He adds, "Olafsson's statements regarding the habits of the birds that breed upon the rocks is in accordance with the relations of other authors. In narrating the capture of the last pair of Garefowls at Eldey in 1844, the writer remarks: [7] "As the men clambered up they saw two Garefowls sitting among numberless other rock-birds (*Uria troile* and *Alca torda*), and at once gave chase. The Garefowls showed not the slightest disposition to repel the invaders, but immediately ran along under the high cliff, their heads erect, their little wings somewhat extended. They uttered no cry of alarm, and moved, with their short steps, about as quickly as a man could walk. Jon (Brandsson), with outstretched arms, drove one into a corner, where

---

[4] A writer in the "English Pilot" for 1794, quoted by Sir Richard Bonnycastle (" Newfoundland in 1842," vol. i. p. 232) states—"There is also another thing to be taken notice of in treating of this coast, that you may know this [bank] by the great quantities of fowls upon the bank, namely, shearwaters, willocks, noddies, gulls, penguins (*i.e.* Great Auks), &c., without any exceptions, which is a mistake, for I have seen all these fowls a hundred leagues off this bank, the penguins (*i.e.* Great Auks) excepted."

[5] Mr. (now Professor) A. Newton on Mr. J. Wolley's " Researches in Iceland respecting the Garefowl." *Ibis*, vol. iii., 1861, p. 393.

[6] E. Olafsson og B. Palsson Reise igj. Island. Soröe, 1772, f. 896, 831.

[7] Professor A. Newton on "Researches in Iceland." *Ibis*, vol. iii., 1861, p. 391. See also p. 21 of this work.

he soon had it fast. Sigurðr (Islefsson) and Ketil pursued the second, and the former seized it close to the edge of the rock, here risen to a precipice some fathoms high, the water being directly below it. Ketil (Ketilson) then returned to the sloping shelf whence the birds had started, and saw an egg lying on the lava slab, which he knew to be a Garefowl's. He took it up, but finding it was broken, put it down again. Whether there was not another egg is uncertain. All this took place in much less time than it takes to tell it."

The Garefowl appears to have been either excessively stupid or insatiable in its appetite. One was caught about 1812 near Papa Westray, Orkney, on the open sea by some fishermen, who enticed it to the side of the boat by holding out a few fish,[8] and then striking it with an oar, either stunned or killed it outright,[9] which, does not appear to be certain, but the result was that the bird was captured. Another instance of the same kind occurred with the specimen caught alive at the entrance to Waterford Harbour in May 1834.[10] There are also several notices of its having been caught with baited lines by vessels crossing the banks of Newfoundland.[11]

The Danish naturalist, Olaus Wormius, gives a figure of this bird. He appears to have drawn it from a living specimen he had obtained from the Färoe Islands, and which lived under his care for several months. The peculiarity of the figure is, that it shows the Garefowl with a white ring round its neck,[12] which most probably indicates the bird was in its spring plumage, and was just throwing off the white feathers that appeared on its throat and neck during winter.[13] (Fig. p. 68, also Note, p. 74.)

In Pennant's "British Zoology," vol. ii., London, 1812, p. 146, is given a figure of the bird swimming. Pennant says: "It lays one egg, which is six inches long, of a white colour; some are irregularly marked with purplish lines crossing each other, others blotched with black, and ferruginous about the thicker end. If the egg is taken away it will not lay another that season. . . . It lays its egg close to the sea-mark, being incapable, by reason of the shortness of its

---

[8] "Transactions of Tyneside Naturalists' Field Club," vol. iv. p. 116.

[9] Dr. Latham, "General History of Birds," vol. x.

[10] Thomson's "Birds of Ireland," vol. iii. p. 238.

[11] Audubon, "Ornithological Biography," 1838, and Edward's "Natural History of Birds," part iii. London, 1750.

[12] "Museum Wormianum seu Historiæ Rerum Rariorum" (Copenhagen), Leyden, 1655, p. 301.

[13] "Proceedings of Scottish Society of Antiquaries," vol. i. N.S., p. 98. See also p. 74 of this work.

wings, to mount higher.[14] . . . The length of the bird to the end of its toes is three feet, that of the bill to the corner of the mouth four inches and a quarter; part of the upper mandible is covered with short black velvet-like feathers; it is

Anser magellanicus, seu Pinguini of Olaus Wormius; from Faröe, 1655. Great Auk (*Alca impennis, Linn.*) Facsimile of original figure reduced in size by one-third.

very strong, compressed, and marked with several furrows that tally both above and below. Between the eyes and the bill, on each side, is a large white spot;

[14] Macaulay's "History of St. Kilda," p. 156.

the rest of the head, the neck, back, tail, and wings are of a glossy black. The wings are so small as to be useless for flight, the length from the tip of the longest quill-feathers to the first joint being only four inches and a quarter."

It is of interest to compare the above with the description given by Dr. Fleming,[14] as he had an opportunity of observing the living bird such as has been afforded to no other British naturalist this century—Length, 3 feet; bill, dorsally, 3 inches; in front of the nostrils, $2\frac{1}{4}$; in the gape, $4\frac{1}{2}$; depth, $1\frac{5}{8}$ inches; 7 ridges in the upper and 11 in the lower mandible; legs black; irides chestnut; margin of the eyelid black; inside the mouth orange; head, back, and neck black—the latter with a brownish tinge; quills dusky; secondaries tipped with white; breast and belly white. In winter the brownish-black of the throat and foreneck is replaced with white. When fed in confinement it holds up its head, expressing its anxiety by shaking its head and neck and uttering a gurgling noise; it dives and swims under water, "even with a long cord attached to its foot, with incredible swiftness."

The recorded occurrences of the observation or capture of the Garefowl and the discovery of its remains appear to limit it to the region north of 48 degrees north latitude in the European area, and 42 degrees in the American; and while in the latter its occurrences are well authenticated south of 70 degrees, in the former it is said to have been found only as far north as the borders of the Arctic Circle. By far the most northerly stations at which the Great Auk bred were those situated on the shores of Iceland and east Greenland. (See Professor Steenstrup's Remarks, Appendix IV.) This circumstance may perhaps be accounted for by the more temperate climate prevailing in the eastern area from the effects of the Gulf Stream.

It is almost certain that the Garefowl has not been met with in high northern latitudes east of Norway [15] or Iceland, though there is an unauthenticated report that one was met with at Spitzbergen.[16] There are statements made in the works of various authors on natural history which might lead to the supposition that the native home of the Garefowl was upon the ice-flows of hyperborean

---

[14] "History of British Animals." Dr. Fleming, 1828. 8vo.

[15] "Videnskabelige Meddelelser," 1855, Nr. 3-7, "Et Bidrag til Geirfuglens," p. 95. It seems very doubtful if the bird Professor Steenstrup refers to, and which was shot in 1848 by Herr Laurenz Brodtkorb, of Wardœ, was really a Garefowl. See for further information *Ibis*, vol. iii., 1861, p. 376. Professor Steenstrup, writing us, 13th April 1885, says: "Certainly not a Garefowl."

[16] "British Ornithology," vol. ii. p. 433. Mr. Selby, from whom this report appears to have first emanated, and which has been copied into other works, stated to Professor A. Newton it was a mistake. *Ibis*, vol. iii., 1861, p. 376.

regions,[17] but these statements seem to have been made under a mistaken impression as to what was the real habitat of this bird.[18]

Even so lately as 1868 a well-known Swiss writer on natural history gave a beautifully executed plate in colour, showing the birds sitting on snow-covered ice of great thickness above an ice-hole, while the background is filled in with mountains of ice. The illustration is given in connection with a paper which is a valuable addition to the literature on the Great Auk.[19]

The Garefowl does not appear to have thriven when removed to situations away from the coast; and Mohr informs us that the peasants of Iceland believed the bird was blind when on land.[20] A writer, who lived two centuries ago, had several at different times, which were easily tamed, but did not live long inland. These birds were caught at the Faröe Isles.[21] Another specimen from the same locality was sent to the Danish naturalist, Olaus Wormius, and from what he states,[22] it appears to have lived only a few months from the time he received it.

In corroboration of the above, we have the instance of the Great Auk captured at the entrance to Waterford Harbour in September 1834, which, when caught, appeared to be starving, as it came to the side of the boat to get food, and for some days after its capture ate greedily of potatoes mashed in milk. Ten days after it was caught, the bird was sold to Mr. Davis, who sent it to Mr. Gough of Horetown, County Wexford. For about three weeks after the bird's arrival at its new home it was not known to eat anything; but fearing it would succumb from want of food, Mr. Gough caused potatoes and milk to be forced down its throat, and from that time it ate voraciously until within a day or two of its death, which occurred a little over four months from the time of its capture. When in Mr. Gough's possession its principal food was trout and other fresh-water fish, which it preferred to fish from the sea. When supplied with food which it was fond of, it often stroked its head with its foot, and also performed this operation on other occasions. It swallowed the fish entire. It was rather fierce, and stood very erect.

---

[17] "Temminck in the Manuel d'Ornithologie" (2de partie, p. 940, 1820). "Annals of the Lyceum of Nat. Hist.," New York, vol. ii., 1828, p. 432. "The Birds of Europe," vol. v. John Gould, London, 1837. Text leaf to Pl. 400. Richardson's "Fauna Borealis, Americana," 1831, &c.

[18] The *Ibis*, vol. iii. p. 15. Professor J. Reinhardt, Copenhagen, on the Birds of Greenland.

[19] M. Victor Fatio, in "Bulletin de la Société Suisse," tome ii. 1ʳᵉ partie. Plate I.

[20] "Forsög til en Islandsk Naturhistorie," &c., ved N. Mohr, Copenhagen, 1786, p. 28.

[21] Debes, "Færoa Reserata," p. 130. Published 1673.

[22] "Museum Wormianum seu Historiæ Rerum Rariorum" (Copenhagen). Leyden, 1655, p. 301.

Dr. Burkitt supplied the following description of this specimen, now in the University Museum, Dublin :—" This bird, a young female, is not in good plumage ;[23] the head, back, wings, legs, and feet are sooty black ; between the bill and eye, on each side of the head, there is a large patch of white, mottled with blackish feathers ; the neck is white, slightly mottled with black ; the front of the body white, the lesser quills tipped with white.[24]

			Inches.
Length (total) ; tail not perfect	. . . . . .	.	29
,,	,, of folded wing (from carpus to point of longest quill)	.	$5\frac{1}{2}$
,,	,, bill from forehead .	. . . .	$3\frac{3}{4}$
,,	., bill from gape or rictus	. . . .	$4\frac{1}{2}$
,,	,, tarsus . .	. . . .	$2\frac{1}{4}$
,,	,, middle toe .	. . . . .	$2\frac{5}{8}$
,,	,, ,, and nail .	. . . .	$3\frac{3}{8}$
,,	,, inner toe .	. . . . .	$2\frac{1}{12}$
,,	,, ,, and nail .	. . . . ..	$2\frac{6}{8}$
,,	,, outer toe .	. . . .	$2\frac{5}{8}$
,,	,, ,, and nail .	. . . .	$3\frac{1}{16}$
,,	,,· tail, which is broken, may have been about	. .	2
Depth (greatest) of bill, exceeding	. . . . .	.	$1\frac{1}{2}$ "

In addition to the foregoing there is the notice by Audubon of the Great Auk caught by the brother of his engraver on the Banks of Newfoundland. He writes as follows :—" The only authentic account of the occurrence of this bird on our coast that I possess, was obtained from Mr. Henry Havell, brother of my engraver, who, when on his passage from New York to England, hooked a Great Auk on the Bank of Newfoundland in extremely boisterous weather. On being hauled on board it was left at liberty on the deck ; it walked very awkwardly, often tumbling over, bit every one within reach of its powerful bill, and refused food of all kinds. After continuing several days on board it was restored to its proper element." [25]

These are the only statements known to us which point to the Garefowl not

---

[23] It is said this bird died while moulting. "Birds of Europe," H. E. Dresser, vol. viii. p. 564.

[24] "Natural History of Ireland," 1851, vol. iii. pp. 238, 239. Also a valuable paper by T. H. Gurney, jun., in the "Zoologist," 1868, 2d series, pp. 1449-1452.

[25] "Orn. Biog.," vol. iv., 1838, p. 316. Mr. R. Champley, in a letter dated 7th February 1885, referring to the bill of the Great Auk, says : "As the number of furrows give strength to the bill in holding or severing fish when taken, the fewer the furrows the weaker the bill. The furrows act as angle iron does to strengthen a vertical iron tube like the Menai Bridge. When the bird caught a fish there would be a great strain on the bill, for the older birds would take and grasp at heavier fish." See remarks on "fish," p. 72.

thriving when in captivity; but if there are any other records in existence that would throw light on this subject, it would be of interest to have them published.

We believe that the only reference to the food of the Great Auk when in a state of freedom is made by O. Fabricius, who says—"The Great Auk fed on *Cottus scorpius*, or the bull-head, and *Cyclopterus lumpus*, or the lump-fish, and *other fishes* of the same size." [26] From the bony nature of the bodies of the fishes he names, we think it most unlikely that they were among the principal foods of the Great Auk, and if any of those birds were so unfortunate as to swallow any of those fishes they might be expected to suffer internal pangs worse than those of hunger. Fabricius must have had some reason for making the statement, and it is possible that on examining the stomachs of some of those birds he may have found bones of the fishes in question. We suspect that the *other fishes* he refers to would be the staple food of the Great Auk, much more than the two he names. His statement would lead to the belief that the size was more important than the variety of the fish, and this appears to be a confirmation of the observations of others that the Great Auk swallowed fish whole. The same writer tells us, " that the stomach of a young bird in grey down (we do not think this can have been a young *Alca impennis*) captured in August contained roseroot (*Sedum rhodiola*) and littoral vegetable matter, but no fishes." [27] This is possible, as the roseroot grows on the ledges and crevices of sea-cliffs and on the boulders on rocky shores almost to within reach of the waves at high water, and the littoral deposits, composed of the seaweeds and other vegetable matter washed up by the waves, would be all within reach of the young sea birds before they launched out upon the ocean on their own account.[28]

Fabricius is the only writer we know of who says he saw the young of the Great Auk, and who gives even the slightest description of it. Unfortunately what he says makes it more than doubtful that the bird he describes was a young Garefowl. He states that the one he refers to was captured during August, *and was covered with grey down*, " pullum vidi, mense Augusto captum lanuginem griseam tantum habentem." If this young bird was captured by Fabricius during

---

[26] " Fauna Groenlandica," p. 82.

[27] " Fauna Groenlandica," p. 82. Professor Steenstrup, in a letter dated 16th March 1885, says, " This young bird of Fabricius has really nothing to do with *Alca impennis*." Also " Et Bidrag til Geirfuglens," Videnskabelige Meddelelser, 1855, p. 41.

[28] Lightfoot, " Flora Scotica," London, 1777, vol. ii. p. 620, says, " The inhabitants of Greenland eat roseroot as a garden stuff."

the month of August, it seems almost certain that it could not be a young Gare-fowl, for the egg of the Garefowl was hatched out by the middle of June, and the young bird would be well feathered by August, and would have very little if any grey down. It is also to be remembered that the Garefowl laid only one egg each year, and it is generally supposed that shortly after the young one was hatched it betook itself to the sea, as when it came from the shell it was fitted for swimming and diving. It is therefore more likely to have been the young of some other large swimming bird that Fabricius mistook for a young Garefowl, and this may account for the vegetable matter found in its stomach, as the usual food of the Garefowl appears to have been fish.

The only immature specimens of the Great Auk that are believed to have been preserved are a specimen at Prague, and the one in the Newcastle-on-Tyne Museum, which probably came from Newfoundland, but which it is supposed may have been sent from Fabricius to Marmaduke Tunstall of Wycliffe, a collector of zoological specimens who lived during the latter half of last century.[29] This collection of Tunstall's subsequently passed into the hands of Mr. George Allan of Blackwell Grange, near Darlington, from whom it received considerable addi-tions, and in 1822 was bought by G. T. Fox, Esq., F.L.S., for the Literary and Philosophical Society of Newcastle-on-Tyne, and now forms part of the Newcastle-on-Tyne Museum. Mr. Allan had written notes on the various birds, but that part which refers to the Great Auk appears to have been penned after receiving information that was mostly erroneous. Extracts from these notes were published by Mr. Fox in 1827 in a "Synopsis of the Newcastle-on-Tyne Museum," 8vo. At p. 92, Mr. Fox refers as follows to the specimen of the Great Auk under con-sideration:—" Our bird is apparently a young one, agreeably to Mr. Allan's remark. I add some description of it, as the young was not known to Temminck. Neck black, spotted or mottled with white; bill, upper mandible, with one large sulcus at the base, none at the tip, in this respect analogous to the young and old razorbill (*Alca Torda* and *Pica*, Linn.); six or eight grooves at the tip of the lower mandible, but without the white ground."[30]

It is somewhat extraordinary that with this exception, and possibly another

---

[29] Latham ("General Synopsis of Birds," London, 1785, 4to. iii. p. 312), says, "In Mr. Tunstall's Museum is one of these (Great Auks), with only two or three furrows on the bill, and the oval space between the bill and the eye speckled black and white. This is probably a young bird." Professor Steenstrup, writing us 13th April 1885, informs us that it is a mistake to suppose that this young bird was sent from Fabricius.

[30] See Appendix, Professor Blasius' remarks, p. 18.

K

skin preserved at Prague in Austria,[31] all the stuffed skins are those of mature birds; but it is perhaps attributable to the desire on the part of the captors to get as fine specimens as possible for collections, as under ordinary circumstances these would be of most value, and the scientific world only wakened up when it was too late to a sense of the want of stuffed specimens to illustrate the various stages of Great Auk life.

The plumage of the bird is known to have varied according to the season of the year, and a similar change takes place among allied species. The following evidence all points to the conclusion that it had more white feathers upon its throat and neck in winter than in summer. In the figure we give of the Great Auk at p. 68, which is a reduced fac-simile of the original picture taken by Olaus Wormius in 1655 from a living bird, the peculiar feature is the white ring round its neck, which we may believe would not been shown there unless it had existed in the live specimen, or had been put on the figure for some purpose. The late Dr. John Alexander Smith, referring to this specimen, remarks this white collar, and thinks it may be explained "*as a mere variety, due to the remains of its winter plumage, when the throat and neck are more or less replaced with white.*"[32] Professor John Fleming, D.D., writing of the specimen he obtained at St. Kilda in 1821, says: "*In winter the brownish-black of the throat and fore-neck is replaced with white.*"[33] Mr. H. E. Dresser gives the following descriptions of the Great Auk, but his statement regarding the number of furrows upon the mandible of the young bird in Newcastle-on-Tyne Museum, which is evidently taken from *Latham*, does not appear to be correct, if we may accept what Mr. Fox says, whom we have just quoted:—

"*Adult in Summer.*—Head, hind neck, throat, and entire upper parts, with the wings and tail, black; secondary feathers tipped with white, and between

---

[31] "Bulletin de la Société Ornithologique Suisse," tome ii. 1re partie, p. 82. Professor Steenstrup, writing us 13th April 1885, remarks, "I think it is not so extraordinary, as nearly all the individuals killed and skinned have been caught on the rocks during the breeding season, and consequently all have been old birds; proportionately very rarely have they been killed on the sea."

[32] "Proceedings of the Society of Antiquaries of Scotland, 1878-79, p. 98.

[33] "Edinburgh Philosophical Journal," vol. x. p. 97. Professor J. Steenstrup, writing us on 4th February 1885, mentions that the white neck ring shown in the figure of the Great Auk in the "Museum Wormianum seu Historiæ Rerum Rariorum, 1655," and of which we give a reduced reproduction at p. 68, is artificial. Professor Steenstrup, 13th April 1885, draws our attention to what he states, "Et Bidrag til Geirfuglens," "Videnskabelige Meddelelser," 1855, No. 3-7, p. 84, note * * *. "An artificial ring perhaps bearing a name or inscription." He adds, "There is no reason why such a ring should be produced by the change of plumage."

the beak and the eye there is also a large oval patch of white; breast and under-parts generally pure white; beak and legs black, the former very strong, and with several vertical furrows on the lower mandible; iris deep brown. Total length, about 30 inches; beak, 3 inches 6 lines; wing, 6 inches; tail, 2 inches; tarsus, 2 inches 1 line.

"*Adult in Winter.*—Figured by Donovan from specimen formerly in the Leverian Museum, as having the chin, throat, and front of the neck white instead of black.

" *Young* (only in the Newcastle Museum). Like the adult, but having only two or three furrows on the mandible instead of from six to ten." [34]

Professor Steenstrup informs the author that one of the stuffed skins in the Royal University Zoological Museum, Copenhagen, is that of a bird in winter plumage.[35]

Edmund de Selys-Longchamps, writing of the specimen preserved in the Dublin Museum, says that of all the many specimens examined by him in the different museums of Europe, this is the only one in winter plumage; [36] possibly he had not seen the Copenhagen specimen.

---

[34] "Birds of Europe." H. E. Dresser, vol. viii. p. 563. The youngest specimen known is that in New-castle; but there is also another young stuffed skin in Prague, pp. 73, 77, and Appendix pp. 17, 21. In a letter, dated 7th February 1885, Mr. R. Champley writes us as follows: "Respecting the immature Newcastle bird, when I saw it a fortnight ago I noticed one furrow at base, and one furrow at extremity of bill. My own has eight furrows at the extremity, and proves the former to be a young bird and the latter an old one." Professor A. Fritsch of Prague has kindly sent us a figure of the young bird in the Prague Museum as it appears in his great work, "Naturgeschichte der Vogel Europa's;" and if the drawing is correct, it must be considerably older than the young specimen at Newcastle, as six furrows are shown on the upper and seven on the under mandible.

[35] In letter dated 25th August 1883. Writing again on 13th April 1885, Professor Steenstrup informs us that in his opinion Benicken's description of the Garefowl in winter plumage is the best. The bird described by Benicken is the bird in winter plumage in the Royal Zoological Museum, Copenhagen. As this informa-tion regarding the importance of Benicken's description has reached us as we are going to press, and his paper is not accessible at such short notice, we must content ourselves by giving the reference to his information regarding the winter plumage. *Ibis*, 1824, p. 887.

[36] "Comptes rendus des Séances de la Société Entomologique de Belgique," 1876, 7th Oct., p. 70.

CHAPTER XI.

EXISTING REMAINS OF THE GAREFOWL.

THE remains of the Garefowl which have been discovered seem to localise the bird within the region we have already indicated, as they have occurred in the shell-mounds of Denmark ; in two similar deposits on the island of Oronsay, one of the Southern Hebrides ; in an ancient kitchen-midden at Keiss, Caithness-shire ; and also at the Whitburn Lizards, county Durham. Some bones were found in Iceland ; in the United States, at Mount Desert, and Crouches Cave in Maine, and in shell-heaps near Ipswich in Massachusetts, bones representing at least seven Garefowls were discovered ; but at Funk Island, off the coast of New-foundland, by far the largest quantity of remains was obtained. We may rest assured that all that have been found are very few in number, in comparison to what at one time existed ; and that there is any good prospect of the further dis-covery of extensive deposits of Garefowl bones may well be doubted.

It has been suggested by the late Dr. J. Alex. Smith[1] that the kitchen-middens of the ancient inhabitants of St. Kilda would likely yield a rich harvest. But we doubt it exceedingly, as Martin[2] mentions that the natives used the entrails and bones of the birds they killed along with other materials for making up a compost to put upon the land. This being the case, the greater part would soon disappear ; and if any bones remain they must evidently be looked for at the time of ploughing the cultivated ground.

It is most difficult to give an absolutely correct list of all the known remains of the Great Auk, as year by year new discoveries are being made of skins, bones, and eggs, which, from various causes, have not hitherto been brought to light. There is also another difficulty, and that is to avoid enumerating the same remains more than once ; as some remains that were mentioned as being in public and private collections not many years ago, have since changed hands, and it is sometimes difficult to discover where they have found a resting-place.

---

[1] " Proceedings of the Scottish Society of Antiquaries," 1878–79, p. 103.
[2] Martin's " Voyage to St. Kilda, 1697," p. 18. Published, London, 1753.

Bearing these things in mind, our readers must be indulgent and not become too severe critics, if they find mistakes in the following lists, which we have prepared from the lists of M. Victor Fatio,[3] and the additions and corrections of those lists made by Professor A. Newton,[4] along with the information given by Professor Wh. Blasius in his recent valuable paper,[5] of which we give partial translations (see Appendix, pp. 4–34), combined with what knowledge we ourselves hae been able to bring to bear upon the subject.

LIST OF REMAINS OF THE GREAT AUK (*Alca impennis*, Linn.)

SKINS.

Country.	Place and Collection.	No.	Authority.
Austria.	Graz. Joanneum. J. W. Clark, in litt. 15th October 1868. See Appendix, p. 12.	1	Prof. A. Newton.
,,	Prague. Part collection Serme ; one of these is supposed to be a young bird. Professor Anton Fritsch has been kind enough to forward to us figures of these specimens as they appear in his great work, " Naturgeschichte der Vögel Europa's." The adult is similar to other mature specimens, and has the plumage of winter or early spring ; but the skin of the young bird is remarkable in that it wants the white mark on its head in front of the eye, and that, instead of the upper part of its neck in front being black, as in the adult, it is grey or speckled, especially at the sides of the neck where the lighter plumage of the front merges upon the dark feathers of the back, which are not, however, so dark as in the old bird. The plumage of the lower part of the body is similar to that of the adult, only a few more dark feathers appear as a patch upon the white plumage immediately below the wings. See Appendix, p. 21.	2	Herr A. Fritsch. Prof. W. Blasius.
,,	Vienna. Imperial Royal Museum. See Appendix, p. 24.	1	Herr W. Passler.
Belgium.	Brussels. Town Museum. See Appendix, p. 8.	1	Prof. A. Newton.
,,	Longchamps. Collection of the Baron de Selys-Longchamps. See Appendix, p. 15.	1	,, ,,
British Isles.	Boyne Court, Essex. Collection of Mrs. Lescher. See Appendix, p. 7.	1	,, ,,
,,	Brighton—Chichester House, East Cliff. Collection of the late Mr. George Dawson Rowley ; now in the possession of his son, Mr. G. Fydell Rowley. See Appendix, p. 8.	2	,, ,,
,,	Cambridge. University Museum. See Appendix, p. 8.	1	Jenyns.
,,	Clungunford House. Aston-on-Clun, Shropshire. Collection of the late Mr. Rocke ; now in the possession of Mrs. Rocke. See Appendix, p. 9.	1	Prof. A. Newton.
,,	Dublin University Museum. This specimen is said to be the only one known in winter plumage, but we think there are probably one or two other specimens with the same feather-	1	Mr. R. Champley.

[3] " Bulletin de la Société Ornithologique Suisse," vol. ii., part 1, pp. 80–85.
[4] *Ibis*, vol. vi., N. S., p. 256.
[5] " Zur Geschichte der Ueberreste von Alca impennis, Linn." Naumburg, a/S. 1884.

SKINS—*continued.*

Country.	Place and Collection.	No.	Authority.
British Isles— *continued.*	ing. See p. 71, Prague, p. 77, Copenhagen, p. 79, also Appendix p. 10.		
	Durham University Museum. See p. 22 and p. 91, also Appendix, p. 10.	1	Mr. R. Champley.
,,	Floors Castle, Roxburghshire. Collection of Duke of Roxburghe, seen by the author. It is not well stuffed, and we think might be improved. See Appendix, p. 11, also p. 92.	1	Mr. J. Gibson.
,,	Hawkstone, Shropshire. Collection of Viscount Hill. See Appendix, p. 13.	1	Mr. R. Champley.
,,	Leeds. Museum of Philosophical Society. See Appendix, p. 14.	1	,, ,, ,,
,,	Leighton. Wales. Collection of Mr. Naylor. See Appendix, p. 14.	1	Prof. A. Newton.
,,	London. Natural History Collection, British Museum, seen by author. The first of these specimens was bought by Dr. Leach at the sale of the effects of Mr. Bullock on 5th May 1819, for £15, 5s. 6d., and deposited in the National Collection. The second was obtained by the Museum in 1856. It came from the collection of Professor Van Lidth de Jeude, who obtained it from the Royal Museum, Copenhagen, to which institution it had come from Iceland subsequent to the year 1830. At one time the word Labrador was marked on its stand. See p. 10, also Appendix, p. 15.	2	Mr. R. Champley.
,,	London. A specimen which belongs to Lord Lilford is at present deposited in the rooms of the Ornithological Union, 6 Tenterden Street ; but will probably be ere long removed to his lordship's seat, Lilford Hall, Oundle, Northamptonshire. See Appendix, p. 15.	1	Prof. A. Newton.
,,	Newcastle-on-Tyne. Museum of the Natural History Society of Northumberland, Durham, and Newcastle-on-Tyne. There are two specimens preserved here. The first of these is a young bird, which has been killed before it had winter plumage, and is a unique specimen in that respect. For further particulars, see p. 73. The other skin is that of an old bird in summer plumage, which until recently belonged to Mr. John Hancock, who has been most handsome in his donations to the Museum. See Appendix, pp. 17 and 18.	2	Mr. R. Champley.
,,	Norwich Town Museum. See Appendix, p. 19.	1	Prof. A. Newton.
,,	Osberton, near Worksop, Nottinghamshire. Collection of Mr. F. W. Foljambe. See Appendix, p. 20.	1	,, ,,
,,	Poltalloch, Lochgilphead, Argyllshire. Collection of Mr. John Malcolm. See p. 94, also Appendix, p. 21.	1	,, ,,
,,	Scarborough, Yorkshire. Collection of Mr. R. Champley. See Appendix, p. 21.	1	Mr. R. Champley.
,,	York. Museum of the Yorkshire Philosophical Society (seen by the author). There are two stuffed skins, but there is a great contrast between the specimens. The one in the Rudston Collection is a very fine skin, carefully covered by a glass shade, which is again placed within the ordinary glass-fronted wall-case. The other skin is in the gallery of the Foreign Bird Room, and belongs to the Strickland Collection, but is in bad condition ; and looks as if it had at one time been greatly exposed to dust. Its feathers are ruffled, or perhaps in some instances broken. We think it might be greatly improved. See Appendix, p. 24.	2	,, ,,
Denmark.	Aalholm, Nysted, Laaland. Collection of Count Raben. The author was informed of the existence of this specimen by Professor J. Steenstrup in a letter dated 25th August 1883. See also Appendix, p. 4.	1	Prof. J. Steenstrup.

SKINS—*continued.*

Country.	Place and Collection.	No.	Authority.
Denmark.	Copenhagen Royal University Zoological Museum. One in summer plumage and the other in winter plumage. Professor J. Steenstrup, in a letter dated 11th September 1884, informs me that the specimen in winter plumage undoubtedly came from Greenland, and is the only known specimen from that locality. See Appendix, pp. 9 and 14.	2	Prof. J. Steenstrup.
France.	Abbeville. Town Museum. See Appendix, p. 5.	1	Prof. A. Newton.
„	Amiens. Town Museum. See Appendix, p. 6. This skin is the property of the town, but is at present under the roof of an old house, where, however, it is in a good state of preservation. It is preserved there along with other stuffed birds.	1	„ „
„	Chalon-sur-Saône. Collection of Dr. B. F. de Montessus. See Appendix, p. 8.	1	Prof. W. Blasius.
„	Dieppe. Collection of Mons. Hardy; understood now to be in the Musée de la Ville. See Appendix, p. 9.	1	Mr. R. Champley.
„	Lille. Musée d'Histoire Naturelle de la Ville. See Appendix, p. 15.	1	Mons. L. Olphe-Gaillard.
„	Paris. Musée d'Histoire Naturelle, Jardin des Plantes. See Appendix, p. 20.	1	Herr. W. Preyer.
„	Paris. The private collection of Mons. Jules Vian. This specimen was recently heard of by Prof. Wh. Blasius of Brunswick, and referred to at a Congress of Natural History Societies held at Magdeburg on 22d September 1884. In a paper read on that occasion, Professor Blasius says : "In the spring of this year (1884), during a journey to Russia, I visited Warsaw, where I heard from Herr Lad. Taczanowski, the keeper of the Zoological Museum in that city, that he had frequently seen a very fine specimen of a stuffed skin of *Alca impennis* in the private collection of Mons. Jules Vian at Paris, and that the last time he saw it was quite recently. This skin has not been recorded in previous lists. See p. 95.	1	Prof. W. Blasius.
„	Vitry-le-François. Collection of the Count de Riscour, who informed Mons. Victor Fatio regarding this skin in a letter dated 9th April 1869. See Appendix, p. 24.	1	Mons. Victor Fatio.
Germany.	Berlin. Royal Zoological Museum. See Appendix, p. 6.	1	Lichtenstein.
„	Bremen. Municipal Collection of Natural History. See Appendix, p. 7.	1	Herr K. Bolle.
„	Breslau. University Zoological Museum. See Appendix, p. 7. Presumably male and female.	2	Herr Alex. von Homeyer.
„	Brunswick. University Museum. Of the two skins here one is only on loan. See Appendix, p. 7.	2	Prof. A. Newton.
„	Darmstadt. There is an imitation specimen here, of which only the head is genuine. See p. 85, and also Appendix, p. 9.	0	
„	Dresden. Royal Zoological Museum. See Appendix, p. 10.	1	Herr W. Preyer.
„	Flensburg. No skins are now known in this town. See Appendix, p. 10.	0	
„	Frankfort-on-Maine. In the Museum of the Senckenberg Society of Natural History. See Appendix, p. 12.	1	Herr Alex. von Homeyer.
„	Gotha. Ducal Museum. See Appendix, p. 12.	1	Dr. Hellmann.
„	Hanover. Provincial Museum. See Appendix, p. 13.	1	Cabanis.
„	Kiel. University Zoological Museum. See Appendix, p. 13.	1	Prof. W. Blasius.
„	Köthen. Anhalt. The Ducal (formerly Naumann's) Collection. See Appendix, p. 13.	1	Herr W. Preyer.
„	Leipzig. University Museum. See Appendix, p. 14.	1	Prof. A. Newton.
„	Mainz. Town Zoological Museum. See Appendix, p. 16.	1	Herr W. Preyer.
„	Metz. Town Museum. See Appendix, p. 16.	1	Prof. A. Newton.

SKINS—*continued.*

Country.	Place and Collection.	No.	Authority.
Germany —*cont.*	Munich. Zoological Museum of the Royal Bavarian Academy of Sciences. See Appendix, p. 16.	2	Herr W. Preyer.
,,	Oldenburg. Grand Ducal Museum of Natural History. See Appendix, p. 19.	1	Cabanis.
,,	Strassburg. Town Museum of Natural History in the Academy. See Appendix, p. 22.	1	Herr W. Preyer.
,,	Stuttgart. Royal Cabinet of Natural History. See Appendix, p. 23.	1	Prof. A. Newton.
Holland.	Amsterdam. Museum of Royal Zoological Society. See Appendix, p. 6.	1	Mr. R. Champley.
,,	Leyden. Zoological Museum. See Appendix, p. 15.	1	Sclater.
Iceland.	Reykjavik. There may be two skins here, though it is doubtful. For the first of these,—See under Flensburg, Appendix, p. 10. This skin is said to be now in the Central Park Museum, New York.	0	Prof. W. Blasius.
,,	The second was mentioned to us in a letter from Mr. R. M. Smith, 4 Bellevue Crescent, Edinburgh, dated 9th December 1884. He informed us that he visited Iceland in 1858, and when in Reykjavik saw at the house of Mr. Olsen a stuffed specimen of the Great Auk, which was said to be the last that was shot by Mr. Siemson, who mentioned the fact to Mr. Smith, and told him that at the time it was killed another Great Auk was left alive at the skerry south-west of Cape Reykjanes. Mr. Smith supposed that this Great Auk had been killed between 1855 and 1858, as when he visited Iceland in the former year he did not hear of it; but this must be a mistake, as the "Researches" of Professor A. Newton and the late Mr. Wolley have made clear. Mr. Smith adds, "So far as my recollection serves me, the specimen I saw was much larger than the one recently shown at the Museum of Science and Art." This remark refers to the Floors specimen, which was brought to Edinburgh to be exhibited at a meeting of the Royal Physical Society (see page 93). On receipt of this information we at once wrote Mr. Smith asking if he had heard of the skin more recently, and on the 11th December he writes, "I had written particularly to the Landfoged at Reykjavik, Mr. Thorsteinson, but he could not obtain any information about the specimen I saw, but I may yet be able to learn something further." Professor A. Newton, to whom we mentioned Mr. Smith's statement, wrote us, on 5th March 1885, as follows:—"I think your correspondent has made some mistake. In 1858 Mr. Wolley and I were for some weeks in Reykjavik inquiring in every direction about the Auk, our inquiries being aided by the kindness of the towns-people. We never heard of such a specimen as you mention, and I think I can almost positively assert that there was not one, even in the whole of Iceland, at the time." Professor Steenstrup writes, on 16th March 1885, regarding this specimen, and states, "he thinks there is some mistake about it." See also Appendix, p. 38.	0 or 1	Mr. R. M. Smith.
Italy.	Florence. Museo Zoologico del R. Istituto di Studi Superiori. As to how this Museum acquired the skin there has been some doubt, though more than one guess has been hazarded. The following extract from a letter, dated 6th October 1884, sent to the author by Professor Enrico H. Giglioli seems to decide the matter:—"The specimen we have mounted is in excellent condition, and is all this Museum possesses of *Alca*	1	Mr. R. Champley.

SKINS—*continued.*

Country.	Place and Collection.	No.	Authority.
Italy— *continued.*	*impennis.* I can add now what I was not able to tell to my friend Professor Blasius, that the specimen of the Great Auk in this Museum was procured through exchange from Professor Sundevall of Stockholm." See Appendix, p. 11.		
,,	Milan. Collection of the late Count Ercole Turati. See Appendix, p. 16. Now acquired for the Public Museum, Milan.	1	Mr. R. Champley.
,,	Pisa. Museo Zoologica del Universita. See Appendix, p. 20.	1	Prof. A. Newton.
,,	Turin. Museo Zoologico del Universita. See Appendix, p. 24.	1	Mr. R. Champley.
,,	Veneria Reale. Private collection of the King. Obtained in 1867 from that of the late Pastor Brehm. See Appendix, p. 24.	1	Prof. A. Newton.
Norway.	Naes, near Arendal. Collection of Herr Nicolai Aall. See Appendix, p. 17.	1	Prof. A. Newton.
Portugal.	Lisbon. Museu Nacional (Secçao Zoologico). See Appendix, p. 15.	1	Prof. A. Newton.
Russia.	St. Petersburg. Zoological Museum of the Imperial Academy of Sciences. See Appendix, p. 21.	1	C. F. Brandt.
Sweden.	Lund. Zoological Museum of the University. See Appendix, p. 15.	1	Prof. A. Newton.
,,	Stockholm. In Zoological section of the National Museum of Natural History. See Appendix, p. 22.	1	,, ,, ,,
Switzerland.	Aarau. Town Museum. See Appendix, p. 5.	1	Dr. C. Michahelles.
,,	Cortaillod. Collection of Captain A. Vouga. See Appendix, p. 9.	1	Mons. Victor Fatio.
,,	Neuchâtel. Museum of Natural History. See Appendix, p. 17.	1	Mons. L. Olphe-Gaillard.
United States.	New York. Central Park Museum of Natural History. See Frontispiece, also Appendix p. 19, also note, p. 11.	1	Mr. R. Champley.
,,	There is said to be in this Museum the faulty skin that belonged at one time to Herr Mechlenburg, see Reykjavik, p. 80, also under Flensburg, Appendix, p. 10.	1 or 0	Prof. A. Newton.
,,	Philadelphia. Academy of Natural Sciences. See Appendix, p. 20.	1	,, ,, ,,
,,	Poughkeepsie. New York State. Vassar College. See Appendix, p. 21.	1	,, ,, ,,
,,	Washington. Smithsonian Institute. See Appendix, p. 24.	2	,, ,, ,,

SUMMARY OF SKINS.

Country.	No.	Country.	No.
Austria	4	Norway	1
Belgium	2	Portugal	1
British Isles	22	Russia	1
Denmark	3	Sweden	2
France	8	Switzerland	3
Germany	20	United States	5 or 6
Holland	2		
Iceland	0 or 1	Total,	79 or 81
Italy	5		

(For figures of stuffed Skins of Great Auk, see Frontispiece, also p. 68.)

(For remarks regarding what should be the attitude given to stuffed Great Auk Skins, see Appendix VI.)

L

SKELETONS.

Country.	Place and Collection.	No.	Authority.
British Isles.	Cambridge. Collection of Professor A. Newton and his brother, Mr. E. Newton. This skeleton was imperfect, and was prepared from a mummy got on Funk Island in 1863 (see on p. 28). It is described by Professor Owen (Transactions of Zoological Society, vol. v. pp. 317–335, pls. li. lii.) More recently it has been made perfect, or nearly perfect, with bones from the extremities of the stuffed skin in the University Museum of Zoology.	1	Prof. A. Newton.
,,	London. British Museum. This skeleton is very perfect. It was obtained from the mummy Great Auk sent from Funk Island in 1864. See pages 6 and 28.	1	,, ,,
,,	London. British Museum, Palaeontological Department, South Kensington. There is in this collection a skeleton which has been recently constructed from bones that belong to the collection made by Professor John Milne at Funk Island. This skeleton is tolerably perfect. See page 100.	1	Mr. Ed. Gerrard, jr.
,,	London. Royal College of Surgeons. This skeleton is of old date, and probably belonged to Mr. John Hunter, from whom the Hunterian Collection derives its name. It has been prepared from a complete body—perhaps from an old dried-up specimen which at one time was the property of the Royal Society. This skeleton is very perfect.	1	Prof. A. Newton.
,,	London. Collection of Lord Lilford. This skeleton was constructed from bones collected at Funk Island in 1874 by Professor John Milne. It is at present at the rooms of the Ornithological Union, 6 Tenterden Street, but in all probability will go from there to the mansion of his lordship, Lilford Hall, Oundle, Northamptonshire.	1	Lord Lilford.
France.	Paris. Museum of Natural History in the Jardin des Plantes. This is a very perfect specimen. It is of old date, and was probably prepared from a fresh corpse from Newfoundland.	1	Prof. A. Newton.
Germany.	Dresden. Royal Zoological Museum. Constructed from bones obtained at Funk Island in 1874, by Professor John Milne.	1	Prof. Wh. Blasius.
Italy.	Milan. Private collection of the late Count Ercole Turati, now in the Public Museum. Constructed from bones collected at Funk Island in 1874, by Professor John Milne.	1	,, ,, ,,
United States.	Boston. Harvard University Museum. These two skeletons were prepared from mummy Great Auks obtained at Funk Island during 1864. See p. 29.	2	Prof. A. Newton.

DETACHED BONES.

Country.	Place and Collection.	No. of Birds represented.	Authority.
British Isles.	Cambridge. Collection of Prof. A. Newton and his brother, Mr. E. Newton. Bones of at least eight individuals, found in kitchen-middens in Iceland by the late Mr. Wolley and Prof. A. Newton. *Ibis*, 1861, pp. 394–396.	8	Prof. A. Newton.
,,	Edinburgh. Museum of Science and Art. Three perfect or nearly perfect bones and four fragments (all representing different bones) that may have belonged to more than one bird. These remains were obtained by the author during the excavations at Caisteal-nan-Gillean, Oronsay, and presented by him to the Museum with the consent of Major-General Sir J. C. M'Neill, V.C. See p. 53.	1	

DETACHED BONES—*continued.*

Country.	Place and Collection.	No. of birds Represented.	Authority.
British Isles—*continued.*	Edinburgh. Museum of Science and Art. Some bones representing possibly more than one individual. The bones were obtained by Professor J. Milne at Funk Island, and we do not count them among the remains here, as they are included in the total of the remains collected by Professor J. Milne at Funk Island in 1874. See p. 85.		
,,	Edinburgh. One fragment of a left humerus at present in the possession of the author. This is the bone of a distinct Great Auk from that mentioned, p. 82 (Edinburgh). See p. 53.	1	
,,	Edinburgh. One bone, perfect or nearly perfect, and six fragments. There is also another bone which is said to have belonged to the Great Auk, but it does not appear, so far as we know, to have been identified with certainty. These bones were all collected during the excavations at Caisteal-nan-Gillean, Oronsay, by Mr. W. Galloway, who exhibited them at the International Fisheries Exhibition, London, 1883. These remains represent at least three specimens.	3	
,,	During part of the summer (1884), Mr. W. Galloway was excavating at a second shell-mound on Oronsay, and in a communication to Mr. Alexander Galletly states that among other remains he has discovered a coracoid bone of *Alca impennis*. It is just possible that when this mound is fully explored more remains of the Great Auk may be found, but at present there is only this bone to record.	1	Mr. Alex. Galletly.
,,	Edinburgh. National Museum of Antiquities. Three perfect or nearly perfect bones, and three fragments. These remains were discovered by the late Mr. Samuel Laing during some excavations at the Birkle Hill Kist, Keiss, Caithness-shire. Vol. i. N.S. "Proceedings Scot. Society of Antiquaries," pp. 78, 79. See also figs. pp. 44, 45. The above bones have belonged to at least two birds. Seen by author.	2	The late Dr. J. Alex. Smith.
,,	London. Royal College of Surgeons Museum. One fragment, the anterior portion of the sternum of a Great Auk, obtained by Mr. (now Dr.) Joseph Anderson (keeper of the Scottish National Museum of Antiquities), at the Harbour Mound, Keiss, Caithness-shire. "Proceedings Scottish Society of Antiquaries," vol. i. N.S. p. 81.	1	The late Dr. J. Alex. Smith.
,,	One cranium, which is believed to be very old. It is supposed to be mentioned by Mr. Nehemiah Grew as far back as 1681. He refers to a dried Penguin, of which this is thought to be the head (Museum Regalis Societatis, London, 1681, pp. 71, 72).	1	Prof. H. Flowers.
,,	Some bones obtained at Funk Island by Professor John Milne in 1874, are preserved in this Museum, but are not here enumerated, as they are included in total at p. 85.		
,,	Mr. Edward Bidwell, Fonnereau House, Twickenham, has recently purchased some of the bones got on Funk Island by Professor Milne that were in the hands of Mr. E. Gerrard, jun., enumerated at p. 85.		Mr. E. Bidwell.
,,	Some beaks, leg bones, &c., that belong to Professor Milne's find at Funk Island are in the possession of Mr. Edward Gerrard, jun., 31 College Place, Camden Town, N.W. They are included in the total at p. 85.		Mr. E. Gerrard, jun.
,,	Some bones of *Alca impennis* are mentioned as having been used to illustrate an Osteological Lecture by a Mr. Blyth For particulars see p. 101.	0 or 1	Prof. A. Newton.

DETACHED BONES—*Continued.*

Country.	Place and Collection.	No. of Birds Represented.	Authority.
British Isles— *continued.*	Newcastle-on-Tyne. Museum of Natural History Society. One upper mandible, discovered among remains from a cave at the Whitburn Lizards, County Durham. See figure, p. 64.	1	Mr. John Hancock.
,,	Two crania and the wing and leg-bones of two birds, are bones that were extracted from the two skins in the Museum by the skill of Mr. J. Hancock.	2	,, ,,
,,	Waddon. In the collection of Mr. Crowley there are some bones recently bought from Mr. Edward Gerrard, jun. They belong to the collection made on Funk Island by Professor J. Milne, enumerated at p. 85.		Mr. E. Bidwell.
,,	York. In the museum are seven bones that were purchased some time ago from Mr. Edward Gerrard, jun., London. They are part of Professor Milne's find on Funk Island, and enumerated at p. 85.		
Denmark.	Most of the bones of the Great Auks that are preserved in Denmark are in the Royal University Zoological Museum, Copenhagen, but a few are in private collections in the provinces; and when we are not quite certain of the places at which the different collections are now to be found, we do not mention any locality. For most of the information regarding the Danish remains we are indebted to Professor J. Steenstrup.		
,,	Bones of two individuals found in an ancient kitchen-midden at Meilgaard, in Jutland. Two right humeri and a radius from the right side of an old bird, slightly injured at the ends. Obtained by Professor J. Steenstrup during 1855.	2	Prof. J. Steenstrup.
,,	One right humerus from an ancient kitchen-midden at Havelse, in Seeland (situated at the southern part of the Issefjord). Obtained by Herr Feddersen during 1856.	1	,, ,,
,,	Bones representing part of the remains of three individuals from an ancient kitchen-midden at Gudumlund, some English miles south of the Limfjord in Jutland, discovered during 1873.	3	,, ,,
,,	Remains representing three individuals from an ancient kitchen-midden at Sölager in Seeland (situated at the northern part of the Issefjord), discovered during 1873.	3	,, ,,
,,	Some bones from Fannerup, not far from Meilgaard in Jutland; but we have been unable to ascertain how many individuals they represent or the year they were got.	1 or 2	,, ,,
,,	One nearly complete cranium, regarding which, on 16th March 1885, Professor Steenstrup has kindly sent us the following information:—"The origin of this Cranium has evidently been misunderstood by Professor Blasius from what he says in his recent paper ('Zur Geschichte der Ueberreste von *Alca impennis, Linn.,* p. 140'). It is of more recent date still, as belonging to the inventarium of the Royal Kunstkammer, it dates back more than a century."	1	
,,	Two crania and other bones representing several individuals, perhaps five or six, sent from Funk Island, off the coast of Newfoundland, by the late Herr P. Stuvitz. See p. 34.	5 or 6	
,,	The wing-bones and metatarsus that were taken from a stuffed skin in the Zoological Museum of the Royal University, Copenhagen. See p. 34.	1	
Faröe.	Island of Sandoe. When Professor J. Steenstrup visited Faröe in 1844, he heard of a Great Auk's head that was preserved on Sandoe, but when Mr. Wolley was at Faröe in 1858 he could	0 or 1	,, ,,

DETACHED BONES—*continued.*

Country.	Place and Collection.	No. of birds Represented.	Authority.
Faröe. *continued.*	find no trace of it. However, we mention the circumstance, as it is just possible it may yet be found. See p. 10.		
France.	Caen, Normandy. There is preserved here an imperfect cranium, which belonged to the find of Professor J. Milne got at Funk Island in 1874. This bone was bought by Professor de Longchamp. It is included in total of remains. See below, Newfoundland.		
Germany.	Berlin. Royal Zoological Museum. There are in this institution the bones of several Great Auks that were brought by Professor J. Milne from Funk Island in 1874. We include them in total. See below, Newfoundland.		
,,	Brunswick. There are preserved in the Ducal Museum of Natural History a number of bones from Funk Island that were bought in London in 1881, and doubtless are some of those found at that island by Professor J. Milne in 1874. These bones may belong to one individual. They are included in the total. See below, Newfoundland.		
,,	Darmstadt. Grand Ducal Cabinet of Natural History. The cranium of the imitation specimen in this collection is genuine. See p. 79, also p. 113.	1	
Newfoundland.	Professor J. Milne, during his visit to Funk Island in 1874, obtained remains that have belonged to about fifty Great Auks, and these bones are now scattered among museums and private collections. In the foregoing lists we have endeavoured to avoid enumerating any of those bones, as we give the total here; but we have mentioned the existence of some of them in several collections. See page 101.	45 or 50	Prof. J. Milne.
Norway.	Christiania. University Museum. Bones of several individuals sent from Funk Island by the late Herr P. Stuvitz.	8 or 10	Prof. A. Newton.
	Bones sent home from Funk Island by Herr P. Stuvitz, but only recorded for the first time during 1884. See p. 100.	20 or 23	Prof. R. Collett.
United States.	Some bones representing at least seven different Great Auks found at Mount Desert and Crouches Cave in Maine, and shell-heaps near Ipswich in Massachusetts. J. Wyman, "Am. Nat." i. pp. 574–578. We have been unable to ascertain where these bones are at present preserved.	7	Prof. A. Newton.

SUMMARY OF SKELETONS.

British Isles   5
France   1
Germany   1
Italy   1
United States   2

Total   .   .   .   10

NUMBER OF GREAT AUKS REPRESENTED BY DETACHED BONES.

British Isles	21 or 22
Denmark	17 or 19
Faröe	0 or 1
France (under Newfoundland) . .	0
Germany (partly under Newfoundland) . .	1
Newfoundland (remains collected by Professor J. Milne, less four skeletons) . .	about 45 or 50
Norway	about 30 or 31
United States	7
	about 121 or 131

DESCRIPTION OF PLATE IX.

All the bones are drawn of natural size.

Figs. 1–5. Dorsal vertebra, shown in its different faces :—(1) side view, (2) from above, (3) from below, (4) in front, (5) from behind.

Figs. 6–8. The entire left coracoid in three aspects :—(6) exterior, (7) interior, (8) its antero-inner edge.

Fig. 9. Upper moiety of right coracoid viewed from the front and inside.

Figs. 10–14. Different views of the right humerus :—(10) posterior surface, (11) anterior surface, (12) external front edge, (13) superior condyloid extremity, (14) inferior condyloid extremity.

Fig. 15. Distal segment of left humerus.

Figs. 16–19. Different views of the distal end of the right tibia :—(16) interior, (17) posterior (18) anterior, and (19) the inferior face.

S.Grieve.

Linn Soc. Journ. Zool Vol. XVI. Pl 9.

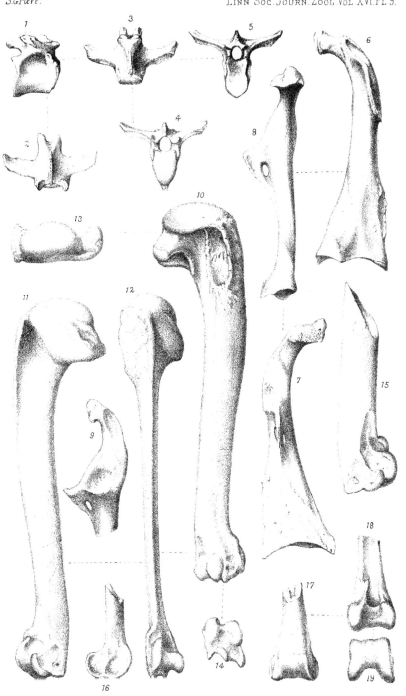

PHYSIOLOGICAL PREPARATIONS.

Country.	Place and Collection.	No. of birds Represented by the Preparations.	Authority.
Denmark.	Copenhagen. Royal University Zoological Museum. Remains of the last two Great Auks killed on Eldey, Iceland, at the beginning of June 1844, preserved as physiological preparations. In a letter dated 25th August 1883, Professor J. Steenstrup says, "To whom the skins of these two individuals, the last killed of all, were sold I do not exactly know. They are male and female (δ $\female$)." In another communication he says, "These specimens were skinned, and the skins sold" (letter of 11th September 1884).	2	Prof. J. Steenstrup.
Italy.	Florence. See p. 97.	0	„ „
Germany.	Hamburg. A specimen in spirit sold about 1840 or 1841, by Herr S. Jacobsen to Herr Selning of this town, or to Mr. Jamrach, the well-known dealer. See p. 102.	1 or 0	Prof. A. Newton.

EGGS.

Country.	Place and Collection.	No.	Authority.
British Isles.	Brighton. The collection of the late Mr. George Dawson Rowley, now in the possession of his son, Mr. G. Fydell Rowley, Chichester House, East Cliff, Brighton. See Appendix, p. 26.	6	Prof. A. Newton and Prof. W. Blasius.
„	Cambridge. Collection of Professor A. Newton and his brother, Mr. E. Newton. See Appendix, p. 26.	3	Prof. A. Newton.
„	Clungunford House, Aston-on-Clun, Shropshire. Collection of the late Mr. Rocke, now in possession of Mrs. Rocke. See Appendix, p. 26.	1	Prof. W. Blasius.
„	Croydon, Surrey. Collection of Mr. Alfred Crowley, Waddon House. See Appendix, p. 27.	1	Prof. W. Blasius.
„	Edinburgh. Museum of Science and Art. The two beautiful eggs in this Museum formed part of an extensive collection of Natural History Objects, which were purchased by some members of the Senatus of the University of Edinburgh in 1819. This collection was afterwards acquired by the Senatus as a body, and in 1855 was transferred by them to the Science and Art Department. The collection was bought principally on account of the stuffed birds it contained. The catalogue which accompanied the collection is complete in most respects, but strangely contains no mention of any eggs. It is entitled "Catalogue des collections d'objets d'Histoire Natlle. formant le Cabinet de Mons. Ls. Dufresne, Naturaliste au Jardin du Roi, Paris, 1815." For further particulars, see p. 107, also coloured plates, p. 108, also Appendix, p. 28.	2	Major H. W. Fielden.
„	Hitchin, Hertfordshire. Collection of Mr. Tuke. See App., p. 28.	1	Mr. R. Champley.
„	Liverpool. The Museum. See Appendix, p. 29.	1	„ „
„	London. Natural History Department, British Museum. See Appendix, p. 29.	2	„ „
„	London. Royal College of Surgeons, Hunterian Museum. See Appendix, p. 29.	3	Prof. A. Newton.
„	London. Collection of Mr. Edward Bidwell of Fonnereau House, Twickenham. This egg has recently been purchased from the sister of the late Rev. George W. Braikenridge of Clevedon, Somerset. For further particulars, see App., pp. 29 and 34.	1	Mr. E. Bidwell.

EGGS—*continued.*

Country.	Place and Collection.	No.	Authority.
British Isles—*continued.*	London. Collection of the late Lord Garvagh, now believed to be in the possession of the Dowager Lady Garvagh. This egg is one of those which his late lordship bought in 1853 from Mr. T. H. Potts, now of Ohinitahi, New Zealand. It is said to have been broken through the carelessness of a footman, and after the death of Lord Garvagh was offered to the late Mr. G. D. Rowley, along with the two whole eggs which that gentleman is said to have purchased from the Dowager Lady Garvagh, but as this specimen was in a broken state, Mr. Rowley declined to buy it. See Appendix, p. 29.	1	Mr. R. Champley.
,,	London. Collection of Lord Lilford. There are here five eggs. The first four of these are mentioned in Appendix at p. 29 ; but the fifth has been purchased quite recently by his lordship in Dorsetshire, and is said to be unrecorded. See Appendix, p. 29.	5	Prof. A. Newton, 1. Mr. R. Gray, 2. Prof. W. Blasius, 1. Mr. E. Bidwell, 1.
,,	London. Collection of Mr. G. L. Russell. See Appendix, p. 30.	1	Prof. A. Newton.
,,	London. There was in the collection of a Mr. Scales an egg referred to by Professor A. Newton, (*Ibis*, 1861, p. 387); but this egg has been lost sight of since 1866. On 4th December 1884 Professor A. Newton wrote to Professor Wh. Blasius of Brunswick, informing him that Mr. Scales died at Brighton in September 1884, aged 90. The son of the deceased informed Professor Newton that the egg was destroyed by fire some twelve years ago. There are plaster casts of this egg in the collections of Professor Newton and Mr. J. Hancock. See Appendix, p. 30.	0	
,,	Newcastle-on-Tyne, in the collection of Mr. John Hancock, which he has recently presented to the Museum of the Natural History Society of Northumberland, Durham, and Newcastle-on-Tyne. See Appendix, p. 30.	1	Mr. R. Champley.
,,	Nunappleton, Yorkshire. Collection of Sir Frederick Milner, Bart. See p. 104 ; also Appendix, p. 31.	1	Prof. W. Blasius.
,,	Oxford, University Museum of Natural History. See Appendix, p. 31.	1	,, ,,
,,	Papplewick, Notts. Collection of Mr. Walter. See Appendix, p. 31.	1	Mr. R. Champley.
,,	Poltalloch, Lochgilphead, Argyleshire. Collection of John Malcolm, Esq. See p. 103, also Appendix, p. 32. Professor Newton informs us that he has possessed a drawing of this egg for some years. In letter 17th September 1884.	1	Prof. A. Newton.
,,	Reigate, Surrey, in the collection of Mrs. Wise (Mr. J. Wolley in MS.) See Appendix, p. 32.	1	Prof. A. Newton.
,,	Scarborough, Yorkshire, Museum of Natural History. See Appendix, p. 33.	1	Mr. R. Champley.
,,	Scarborough, Yorkshire. Collection of Mr. R. Champley. See Appendix, p. 33.	9	,, ,,
,,	Wavendon Rectory, by Woburn, Bedfordshire. Collection of Rev. Henry Burney. This is a fine and perfect specimen. See Appendix, p. 34.	1	Prof. A. Newton.
Denmark.	Copenhagen. Royal University Zoological Museum. In a letter dated 4th February 1885, Professor J. Steenstrup informs us that they have had a splendid figure of this egg drawn by a very good artist, a Mr. Thornam. See Appendix, p. 28.	1	Prof. J. Steenstrup.
France.	Angers. Collection of Count de Baracé. See Appendix, p. 25, see p. 104.	3	Mr. R. Champley.
,,	Angers. Musée de la Ville. See Appendix, p. 25.	1	Prof. Wh. Blasius.

EGGS—*continued.*

Country.	Place and Collection.	No.	Authority.
France—*continued.*	Bergues-les-Dunkerque. Collection of M. De Meezemaker. See Appendix, p. 25. On the 1st October 1861, Mons. De Meezemaker wrote Mr. R. Champley as follows :—"These eggs were brought, in what year I cannot find out, by the captain of a whaling vessel, who gave them in a present to a merchant of this town (Bergues), who gave them to a young man who was commencing a collection of eggs, which I acquired after his death."	2	Mons. Leon Olphe-Gaillard.
,,	Dieppe. Collection de M. Hardy in the Musée de la Ville. See Appendix, p. 27.	1	Mr. R. Champley.
,,	Manonville, Meurthe et Moselle. Collection of Baron Louis d'Hamonville. See Appendix, p. 30.	1	Prof. Wh. Blasius.
,,	Paris. Museum of Natural History, Jardin des Plantes. See Appendix, p. 31. At a meeting of the German Natural History Societies, held at Magdeburg on 22d September 1884, Professor Wh. Blasius mentioned that two eggs of *Alca impennis*, that were previously stated to be in the Natural History Museum at Paris, did not appear to be there now, as Herr Berger, merchant at Witten, had written him saying that he was there this summer (1884), and that he did not find them in the Museum, and that they were unknown to the officials. It is probable they may be in the Lyceum at Versailles, where they were known to be originally.	1 or 3	M. Des Murs, 1. Prof. A. Newton, 2.
,,	Versailles. See Paris. See Appendix, p. 32.	2 or 0	
Germany.	Breslau. Collection of Count Rödern. See Appendix, p. 25.	1	Mr. R. Champley.
,,	Dresden. Royal Zoological Museum. See Appendix, p. 27.	1	,, ,, ,,
,,	Düsseldorf. Museum Löbbeckeanum. See Appendix, p. 27.	1	Prof. Wh. Blasius.
,,	Oldenburg. Grand-Ducal Museum of Natural History. See Appendix, p. 31.	1	Cabanis.
Holland.	Amsterdam. Museum of Zoological Society. See Ap., p. 24.	1	Mr. R. Champley.

,,	Leyden. Zoological Museum. See Appendix, p. 29. The following letter referring to this egg is of interest :—
	
	"LEYDEN, 29*th February* 1860.
	"DEAR SIR,—According to your request, I have the pleasure of sending you here enclosed a drawing (natural size) of the egg of the Great Auk (*Alca impennis*), being for the moment the only specimen in the possession of our Museum. The second one was, a few months ago, presented to the Royal Zoological Society at Amsterdam. Both eggs were procured from a French whaler in the beginning of the century.
	"*To* R. CHAMPLEY, Esq., "*From* H. SCHLEGEL,
	        Scarborough. Director of the Royal Museum
	                    of Natural History of the
	                    Netherlands."
		1	,, ,, ,,

New Zealand.	Ohinitahi, Canterbury. Collection of Mr. T. H. Potts. See Appendix, p. 31.	1	Mr. T. H. Potts.
Portugal.	Lisbon. The Museu Nacional (Secçao Zoologico). See Ap., p. 29.	1	Mr. Ph. L. Sclater.
Russia.	St. Petersburg. See p. 110.	2 or 0	
Switzerland.	Lausanne. Museum of Natural History. See Appendix, p. 28.	1	M. Victor Fatio.
United States.	Philadelphia. Academy of Natural Sciences. There is certainly one egg preserved in this Institution, but it is just possible there may be two. See Appendix, p. 32.	1	Prof. A. Newton.
,,	Washington. Smithsonian Institution. See Appendix, p. 34, and (under Philadelphia) p. 32.	1	Prof. A. Newton.

M

SUMMARY OF EGGS.

Country.	No.	Country.	No.
British Isles	45	Portugal	1
Denmark	1	Russia	0 or 2
France	11	Switzerland	1
Germany	4	United States	2
Holland	2		
New Zealand	1	Total	68 or 70

As far as we have been able to ascertain, the known remains of the Gare-fowl may be totalled as follows :—

Country.	Skins.	Physiological Preparations.	Skeletons.	No. of Birds Represented by Detached Bones.	Eggs.
Austria	4	...	...	...	...
Belgium	2	...	...	...	...
British Isles	22	...	5	21 or 22	45
Denmark	3	2	...	17 or 19	1
Faröe	...	...	...	0 or 1	...
France	8	...	1	...	11
Germany	20	0 or 1	1	1	4
Holland	2	...	...	...	2
Iceland	0 or 1	...	...	...	...
Italy	5	...	1	...	...
Newfoundland (unrecorded remains collected by Prof. J. Milne)	...	...	...	about 45 or 50	...
New Zealand	...	...	...	...	1
Norway	1	...	...	30 or 31	...
Portugal	1	...	...	...	1
Russia	1	...	...	...	0 or 2
Sweden	2	...	...	...	...
Switzerland	3	...	...	...	1
United States	5 or 6	...	2	7	2
Total	79 or 81	2 or 3	10	about 121 or 131	68 or 70

REMARKS ON THE REMAINS.

That there is little hope among naturalists that any great quantity of Garefowl remains may yet be discovered is clearly evidenced by the increasing value put upon its skins, bones, and eggs, but especially the skins and eggs, and so great is the desire to obtain these that unprincipled persons have been known to imitate them. Some of the bones brought from Funk Island were sold at comparatively moderate prices, but probably the British remains, if these were for sale, would fetch high prices, owing to their great rarity.

REMARKS ON SKINS AND THEIR VALUE.

During the last sixty or seventy years many sales of skins have taken place, but as most of these have been carried through privately, it is only in a few instances that the prices given have been made public. It is, however, interesting to observe the rapid increase which has taken place in the value of such remains, and for the purpose of illustrating this we may state that on the 5th May 1819 the skin that was obtained in 1812 from Papa Westra, one of the Orkney Islands, and which had a special interest as being of British origin, was sold at Mr. Bullock's sale for £15, 5s. 6d. In 1832, the specimen now at Neuchatel was bought at Mannheim for 200 francs, or about £8. Two specimens now at Munich were purchased in 1833 for 200 florins (£16, 9s. 2d.) and 50 florins (£4, 2s. 3d.) The skin in the University Museum, Durham, was bought in 1834 or 1835 by the Rev. T. Gisborne from Mr. H. Reid, Doncaster, for £7 or £8 (see Mr. Proctor's letter, page 22). There is now at Gotha, a skin that was purchased in 1835 for 20 thalers from Frank, a dealer in zoological wares at Leipzig. The exact date at which the Poltalloch specimen was bought is unknown, but was probably about 1840; and it was purchased in London from the elder Mr. Leadbetter, dealer in natural history specimens, for a price which, as far as Mr Malcolm, the purchaser, recollects, was £2 or £3.

A skin now in Aarau, Switzerland, was purchased in 1842 or 1843 for 80 florins (£6, 11s. 8d.) Another skin now in Bremen was bought in 1844 for £6. The last two specimens of the Garefowl that were killed on Eldey at the beginning of June 1844, were sold before they reached Reykjavik for eighty rigsbank-dollars, about £9, or £4, 10s. each.

Sixteen years now elapse without the price paid for any skin becoming known to the public, but in 1860 the skin now at Clungunford House, Aston-on-Clun, Shropshire, was obtained by a dealer in natural history specimens from the Museum of Mainz in exchange for the skin of an Indian tapir. About the same time, probably during the same year, the late Herr Mechlenburg of Flensburg sold to Mr. R. Champley of Scarborough a skin and egg, which are now in his collection, and for which he paid £45.

It is stated that Herr Mechlenburg also sold a skin without feet, and from which the breast plumage was awanting, for 1000 marks Schleswig-Holstein currency (= £60), to Siemsen, a merchant in Reykjavik, but this is probably a mistake.[6]   It was through the agency of the same Siemsen that numbers of the Icelandic skins had been forwarded at an earlier date to Denmark.   The specimen figured (see frontispiece), which is now preserved in the American Museum of Natural History, Central Park, New York, and presented to that institution by Mr. Robert L. Stuart, was purchased in 1868 for 625 dollars gold, which, calculating the value of each dollar at 4s. 2d. sterling, gives a total value of £130, 4s. 2d.   This skin belonged to the late Dr. Troughton, who, from a remark in a letter (see p. 106), appears to have bought it from Mr. Bartlett about 1851, and after his demise it was sold by his executors to Mr. Cook, a dealer in natural history specimens, for £91.   Mr. Cook stuffed the skin, and then sold it for £120 to Mr. D. G. Elliot, New York, from whom it was bought by Mr. Stuart and presented as above.   In 1869 one of the skins now in the collection of the late Mr. G. D. Rowley, was purchased by the late Mr. G. A. Frank of Amsterdam for ten louis d'or.[7]

On the 13th April 1870, Mr. Edward Gerrard, junior, wrote to the authorities of the Museum of Science and Art, Edinburgh, offering to sell a skin of *Alca impennis* for £100.   He said that he had seen Mr. Bryce Wright, dealer, in London the day previous, and he wanted £110 for a large skin of a Great Auk which was in fair preservation; that before leaving he (Mr. Gerrard) got Mr. Wright to agree to part with it to him for £100.   The Museum authorities, after consideration, determined to decline the offer, as at that time it was thought to be too high a price to pay.   In the same letter is mentioned the price paid by Mr. Cook for the skin now in New York.   Mr. Gerrard also says that until shortly prior to this time the value of the skins was £80 to £90.

A number of skins are in private collections, where they are quite lost to students of natural history.   If this should meet the eye of the possessor of one of these, perhaps they will permit us to lay before them the claims of the Scottish National Museum, Chambers Street, Edinburgh, which, with one of the finest collections of birds in the three kingdoms, is still without a Great Auk or Garefowl.

It is probable that a few skins as yet unrecorded exist in private collections, and we may instance the two recently brought to light in Scotland, which are the only ones known to be north of the Tweed.   The first of these was found in the Museum of His Grace the Duke of Roxburghe at Floors Castle, Kelso, and its discovery in that collection a few years ago created much interest among

---

[6] This faulty specimen is said to be now in the Central Park Museum, New York (see p. 81).

[7] Equal to about £9, 18s. 5d.   The price seems too small for a skin.

Scottish ornithologists. It is unfortunately not a very good specimen, and the appearance of the skin as at present mounted is apt to convey the impression that the bird had been half starved; but it is probable this is rather due to the taxidermist who stuffed it than to the skin itself, and we venture to express the opinion that in skilful hands the defect might be remedied. Until about 1880 this specimen was only known to a very few individuals, but about that time its existence came to the knowledge of Mr. John Gibson of the Museum of Science and Art, Edinburgh, who succeeded in obtaining the permission of the Duke of Roxburghe to have it recorded. For that purpose it was brought to Edinburgh in April 1883, and exhibited at a meeting of the Royal Physical Society, held on the 18th of that month, when a short paper upon it was read by Mr. Gibson,[8] who thinks it probable that this specimen came from Iceland some time between 1830 and 1840, as it was during that period His Grace the late Duke made the Floors Castle collection of birds, and it is also understood that he visited Iceland. It appears, however, to be very uncertain how the skin came into the collection; and at our request Mr. Andrew Brotherston, curator of the Floors Castle museum, inquired of His Grace the present Duke, and the following is a short extract from a letter on the subject received from Mr. Brotherston, dated 20th June 1882:—"I was at Floors to-day, and saw His Grace the Duke of Roxburghe. He is uncertain about the history of the Great Auk, but has an impression that his father bought it in Edinburgh." In answer to more recent inquiries, Mr. Brotherston writes us, under date 9th September 1884:—"I am sorry to say that I have not been able to learn anything more about the Great Auk at Floors." Professor A. Newton has informed Professor W. Blasius that he conjectures this skin was bought from a London dealer (see Appendix, p. 11).

We believe that it was Mr. Brotherston who was the first to call attention to this specimen, and it was from him that Mr. Gibson heard of it. It is unfortunate that so little is known of its history. The only words written on its stand are —"Great Auk—Male."

Mr. Gibson, in his paper referred to, says—"The bird is an adult male in full summer plumage—the following being its principal dimensions:—

		Inches.
Length from tip of bill to end of tail	. . . . .	$34\frac{1}{2}$
„ of tail .	. . . .	3
„ of tarsus	. . . .	$2\frac{1}{2}$
„ of wing	. . . .	$6\frac{5}{8}$
„ of bill dorsally	. . . . . .	$3\frac{7}{8}$

[8] "Proceedings Royal Physical Society, Edinburgh," 1882–83, pp. 335–338.

									Inches.
Length of gape	.	.	.	.	.	.	.	.	$4\frac{1}{2}$
„ of bill from nostrils	.	.	.	.	.	.			$2\frac{3}{4}$
„ Depth of bill	.	.	.	.	.	.	.		$1\frac{3}{4}$

" There are seven ridges on the upper, and eleven (two of them indistinct) on the lower mandible. The number of ridges and the large size of the bird point to the specimen as being that of an old male."

The second skin that has recently been brought to light in Scotland was, as far as we are aware, recorded for the first time by Professor Wh. Blasius in his work, " Zur Geschichte der Ueberreste von *Alca impennis*," published in the beginning of 1884. He got the information regarding its existence from Professor A. Newton, who, in a letter dated 17th September 1884, informs us that he heard of this skin, and also the egg in the same collection, from Mr. Whiteley, who stated that he had seen them thirty years ago. This skin and egg are preserved at Poltalloch, near Lochgilphead, Argyleshire, and belong to John Malcolm, Esq., who kindly writes us, under date 19th August 1884—"I am sorry I cannot give you much information relative to the specimen of the Great Auk in my possession. I have a very fine and perfect specimen of that bird in my small collection, and also the egg; both these were purchased by me many years ago in London, and I do not know where they came from, but believe they were brought home in one of the Arctic expeditions." In answer to our further inquiries, Mr. Malcolm wrote us on 23d August 1884—"I think you are correct that the specimen of the Great Auk which I have was bought from the elder Leadbetter, but I cannot recollect the price I paid for it, and have no record to refer to of the purchase. . . . If I remember *right*, the specimen of the bird itself did not cost me more than two or three pounds, but, as I before said, I cannot be sure of this, and have no memoranda."

The information regarding this skin and egg did not reach Professor Blasius until the part of his work which refers to the skins was published, so he mentions it in connection with the egg, and he states that Professor Newton thinks these specimens were bought from forty to fifty years ago (see Appendix, p. 32).

Another instance of the recent discovery of a skin has occurred in Denmark, where one has been brought to light in the collection of Count Raben at Aalholm, Nysted, Laaland. The first information we got about this skin was in a letter from Professor J. Steenstrup of Copenhagen, dated 25th August 1883.

We understand that he only became aware of its existence shortly previous to that time (for further information, see Appendix, p. 4).

Since the above discovery was intimated another skin has been brought into notice through the inquiries of Professor W. Blasius, and he recorded this skin in a paper he read on the 22d September 1884, at a meeting of the German Natural History Societies, that met together at Magdeburg. The learned Professor stated that in the spring of 1884, during a journey to Russia, he visited Warsaw, where he heard from Herr Lad Taczanowski, the keeper of the Zoological Museum in that city, that he had frequently seen a very fine specimen of a stuffed skin of *Alca impennis* in the private collection of Mons. Jules Vian at Paris, and that the last time he saw it was quite recently. On the 2d February 1885, Professor W. Blasius kindly sent us the following additional information :—" The specimen was bought at a low price some three years ago from a sailor at Brest, in whose family it had been kept for some fifty years. This skin is also mentioned on the authority of E. Fairmaire, by Olphe-Galliard in his ' Contributions à la fauna ornithologique de l'Europe occidentale.' Fascicule I., 1884, p. 26. Doubts may, however, be cast on the existence of the two birds mentioned by the same author (*ibid.*, p. 29), as in the Collection Dufresne, Paris."

There was at one time a stuffed skin of *Alca impennis* in the Natural History Museum in Manchester, which is now in connection with " The Owens College." On the 9th October 1884, Professor W. Boyd Dawkins kindly wrote as follows—" I am sorry that our specimen of *Alca impennis* disappeared before 1869, and I have never been able to trace it." Immediately on receipt of this letter we wrote thanking the learned Professor for the information, and asking him if he could favour us with any particulars regarding the history of this skin, and also of an egg mentioned as being in the Manchester Museum by Professor W. Preyer of Jena, in 1865. Professor W. Boyd Dawkins, when our letter reached him, was kind enough to hand it to Mr. J. R. Hardy, who, writing from " The Owens College " on the 29th October 1884, says—" Professor Dawkins handed me your letter some days ago. I have had a good hunt, but fail to find any trace of Great Auk or egg. I remember when quite a little lad my father buying the skin and egg for £13, but I cannot just now find the letters. I have gone over thousands of my late father's letters, but so far fail to find the ones regarding the Great Auk."

It is just possible this skin and egg may yet be found. We are not aware that the skin has been previously recorded.

At least one skin that had become decayed has been destroyed, and those who permitted this must surely have been wanting in knowledge of its value.

The skin referred to was in the Teyler Collection in the Haarlem Museum; and Mr. G. A. Frank of London informed Professor W. Blasius that it was purposely destroyed during the management of Professor Breda, and that no bones were obtained from it.

In a letter to Mr. R. Champley, which was probably written about 1860, though it bears no date, Herr Friedrich Schultz, Dresden, informs him that in the year 1835, on March 21st, he received from Mr. F. G. W. Brandt, Hamburg, three skins of *Alca impennis* and two eggs. On the 9th May of the same year he bought personally at Hamburg, from a stranger, two skins and two eggs. He received, on 7th July, from Mr. Salmin, Hamburg, one skin. From that time he did not get any skins of *Alca impennis*. He says, "According to my books these skins and eggs were sold only to the following :—Mr. Henry Ross, Leipzig, one skin; after his death the collection went to Professor Dr. Schweageichen, and later to the Museum of his University; Mr. Robert Draak, Amsterdam, got two skins; Mr. Hugh Reid, Doncaster, two skins and two eggs; Barber Hühnel, Leipzig, one egg. What has become of the rest I cannot state with certainty." Before leaving this part of our subject, it may be as well that we should refer to the number of skins obtained from Iceland between the years 1830 and 1844. It is most difficult to get accurate information upon this matter, and Professor Newton, Professor Preyer, and Professor Blasius all appear to differ, more or less, as to the number of birds captured. The fact seems to be that so many years had elapsed after the last specimens of *Alca impennis* were obtained, before inquiries were instituted as to the actual numbers killed, that those who made raids on Eldey had forgotten the exact numbers; and the statements of Herr Siemsen of Reykjavik, into whose hands a number of the skins and eggs had passed, appear to Professor Newton to be inaccurate. The same writer states, that besides this he has no doubt a number of expeditions to the skerry took place between 1830 and 1844, of which, when he wrote his paper on the late Mr. Wolley's researches in 1861, he could give neither the dates or the results. He then alludes to the total number of skins got during the fourteen years referred to, and says, "If all the stories we received can be credited, the whole number would reach eighty-seven. I should imagine sixty to be about the real amount." [9] Professor Blasius, who has studied the subject carefully, appears to think that probably the

---

[9] Mr. J. Wolley's "Researches," by Prof. A. Newton. *Ibis*, October, 1861, p. 392.

following are the numbers of skins that reached Britain and the Continent of Europe, between 1830 and 1840 during the different years mentioned:—1830, twelve to twenty-one?; 1831, twenty-four; 1833, three to thirteen?; 1834, nine; 1840, three. Total, fifty-one or seventy.[10] To these there has to be added the two birds killed in 1844, the last Garefowls that were seen. Most if not all the Gare-fowls obtained during these years appear to have been skinned by opening them under the right wing, and then the skins were stuffed with hay, the bones being wrapped round with hemp (*Ibis*, 1861, p. 390). The greater number of the skins enumerated by Professor Blasius were sent from Iceland to Copenhagen, but a con-siderable number reached London, and a few went to Hamburg and Flensburg.

SKELETONS.

The two incomplete skeletons said to exist at Breslau and Florence, men-tioned by Mr. R. Champley ("Ann. and Mag. Natural Hist.," 1864, vol. xiv. p. 236), appear to be non-existent, as Professor W. Blasius has made inquiries and can find no trace of them. He wrote Professor Anton Schneider, director of the University Zoological Museum at Breslau, regarding all the remains of *Alca impennis* known to exist there, and requested him for information. Professor Schneider wrote regarding other remains of the Great Auk at Breslau, but he made no mention of a skeleton.

As it was desirable that this point should be settled, he wrote to Professor Schneider, and in a letter dated 3d October 1884, he kindly informed us that "no skeleton or part of a skeleton of the *Alca impennis* exists in either the Zoological or Anatomical Museums of the University here." On the 25th Septem-ber 1884, we received a letter from Mr. R. Champley, in answer to our inquiries as to how he came to make the statement referred to, and apparently there must be some mistake as to Mr. Champley having been the authority for the Breslau skeleton; he says, "I have no recollection of any skeleton at Breslau. I am not aware that my name is mentioned in connection with any skeleton supposed to be in Breslau. If my name has been inserted, it is an error."

As to the incomplete skeleton at Florence, Professor Enrico H. Giglioli wrote Professor W. Blasius, saying, "Not even a bone exists in Florence." This statement seems very clear, but as we thought Mr. Champley might be able to

---

[10] Zur Geschichte der Ueberreste von *Alca impennis*, Linn., Prof. Wilh. Blasius. Naumburg a/S, 1884. p. 126

throw some light upon the subject, and at least state how he came to mention the existence of this skeleton, we wrote him asking for information; and on the 23d September 1884 he wrote, " As regards the skeleton, it was part of one in spirits, and shown me by the curator in his private room in 1861, when I was at Florence." On the 26th September he wrote us further upon the same subject. He says, " I made no note of the results of my inspection of the Museum (the Anatomical and Natural History one), but to the best of my belief the skeleton had a portion of the viscera attached, and was contained in a sealed glass jar about 18 inches by 8 inches. I remember the visit perfectly well." It occurred to us that as there was a considerable difference between an incomplete skeleton of a Great Auk and part of the body of one of these birds preserved in spirits, that it might be well to write and explain this to Professor Enrico H. Giglioli, and ask if he could give any information upon the subject. We accordingly wrote, and the learned Professor kindly replied on 6th October 1884, as follows :—" From renewed inquiries I find that *no skeleton or single bone* of the Great Auk ever was in this Museum, and I cannot imagine how Mr. Champley came to make his statement."

We communicated this information to Mr. Champley, who wrote us on the 11th October :—" I will look among my papers ; I think I have the card of the curator, or at least his name. It is now twenty-three years since. The glass-case contained other birds, only I remember there was a door on each side, and the glass sealed bottle was on a shelf in the centre. As I said before, I took little interest save in the skins and eggs at that time." On the 15th October Mr. Champley again wrote :—" I remember well what was shown me, because at the time I was so much impressed, and I had no object whatever in stating what I then noticed. During my tour in 1861 I preserved all my hotel bills, and the cards, addresses, introductions, &c., are all together somewhere ; I will look them out." On the 18th October he wrote as follows :—" I have looked over my diary, and only find a pencil note similar to my previous statement. I cannot find any name of the curator, but I think his name had three syllables, and there was, I think, a tautology in the pronunciation. I have an indistinct remembrance of the name. Possibly if you had half a dozen names of that date (1861), I could then recall the name." We wrote to Professor Enrico H. Giglioli on the 20th October informing him of what Mr. Champley said, and on the 2d December he answered us as follows :—" I have been absent from Florence for the last month, and that accounts for my not having answered your last. The persons in charge of the Florence Natural History Museum in 1861 were Messrs. Bruscoli and Piccioli; both are alive, and both tell me that they never saw anything like the specimen described by Mr. Champley. There must be some mistake, and I am inclined to think that Mr. C. saw what he describes in some other museum, for even if the supposed specimen had since disappeared from this museum it would be yet on the old catalogues, which I have, and in which I find no mention of such a specimen." When this letter reached us we at once communicated its contents to Mr. Champley, who replied as follows on the 8th December :—" Referring to your letter, I can only add that I made the pencil entry at the time, but neither of the names you give resembles the one I indistinctly remember. At all events, the Great Auk remains are not there. Some museums effect exchanges, and these remains may possibly turn up somewhere, and the mystery be cleared up." On the 13th December 1884, Mr. Champley kindly sent us the original note, which he had found among his papers, and of which the following is a copy. It is written on a scrap of paper, and Mr. Champley says he does not know the name of the person who wrote it, but it was written for him at his request when he was in the Museum at Florence. This note, though it specially refers to a skin, seems to decide the question at issue :—

" Le seul individu de *l'Alca impennis* existant dans la Collection ornithologique du Rl. Musée de Physique et d'Histoire naturelle de Florence, fut acheté dans l'an 1837 du marchand naturaliste Etienne Moricaud de Genève.

" FEDERIGO BRUSCOLI, *Conservateur.*

" FERDINAND PICCIOLI, *Aide du Professeur de
Zoologie des Vertébrés.*

" FLORENCE, *le* 21 *Mai,* 1861."

In a letter, dated 17th January 1885, Mr. Champley says, " I have been thinking over and over again about the viscera of the *Alca impennis* at Florence. It was placed in a glass case fronting a circular table on which stood, I believe, a portion of the hands of Galileo under a shade." If these remains exist anywhere, and they are ever discovered, it will be most interesting, as the only physiological preparations of *Alca impennis* known are those in the Museum at Copenhagen, though the body of another Great Auk is said to have been sent from Iceland in spirits. See p. 103.

As our readers are aware, the remains that were obtained from Funk Island in 1874 by Professor J. Milne are now scattered, and as it was, if possible, desirable that fuller information should be obtained regarding these bones, we communicated with Mr. Edward Gerrard, junr., who kindly answered us as follows on 23d September 1884: "I do not remember the exact number of beaks of the Great Auk, but there were over fifty. I sorted and fitted the bones together as well as I could to make three skeletons, but it is impossible to say of how many different birds each skeleton was made up. Besides the three skeletons you name, there is now another in the British Museum. I sold a series of bones to Dr. Woodward, and since the collection has been at Kensington he has had them put together, and they make a tolerably perfect skeleton. I sold one skeleton to Count Turati, one to Mr. T. C. Eyton, and another to Dr. Meyer. Imperfect heads and odd bones I sold to a good many people." In a postscript Mr. Gerrard informs us that he has still a few beaks, leg-bones, &c., to dispose of, and any persons desiring to obtain Alcine remains would do well to get further information from him. His address is 31 College Place, Camden Town, London, N.W.

The skeleton sold to Count Ercole Turati was still, in 1881, in the collection of that deceased nobleman, as mentioned by T. Salvadori (*Ibis*, 1881, p. 609); but we have recently received information that it is now in the Public Museum, Milan. The skeleton sold to Mr. T. C. Eyton is now in Lord Lilford's collection, and the unmounted skeleton sold by Mr. Gerrard to Dr. A. B. Meyer has been put together and is in the Museum at Dresden.

Detached Bones.

From an examination of the bones sent home by P. Stuvitz from Funk Island, Professor R. Collett has arrived at the conclusion that they represent thirty-six individuals, and he has published the results of his investigations in "Mittheilungen des Ornithologischen Vereins in Wien," 1884, Nos. 5, 6. We are indebted to Professor Steenstrup for this information, which he sent us on 16th March 1885. Unfortunately we have had no opportunity of seeing the papers in question, but if we deduct, from the thirty-six enumerated by Professor Collett, the bones sent by Stuvitz, now in Copenhagen, said to represent five or six specimens, and those in Christiania, said to belong to eight or ten individuals, we have bones representing either twenty or twenty-three Great Auks still to record. (See page 85.)

The bones that have been discovered in the Danish Kjökkenmöddings have not yet been properly recorded, and to a foreigner living in a distant country, and without a knowledge of the language, the difficulties in connection with this work may be looked upon as insurmountable. If some Danish archæologist would undertake this labour, he would be conferring a favour upon all interested in *Alca impennis.* With the desire of tracing to their present resting-places at least some of these remains, we wrote to Professor J. Steenstrup, who has always been so willing to give us information, and he answered our inquiry in a letter dated 25th November 1883, as follows :—"I am very sorry that I am unable to give you such references as in this case I wish to do, because I can by no means spare the necessary time for researches and correspondence with different persons still living in Jutland, and in whose small collections of flint and bone implements, and of bones of mammals and birds, I have observed some relics of *Alca impennis.* Such relics are indeed very few, and it is to be hoped that they will find their way some day to the Zoological Museum in Copenhagen. Of course I must confess that not all the bones of *Alca impennis* found to this day, and known to me, are in the Museum. The number of individuals to which our bones belong I cannot say before they have been more exactly compared, and the relations of tarsal and leg bones to humeri and ulnae, &c., duly considered."

Most of the bones of *Alca impennis* that have been known to exist in Britain have been accounted for, but there are amissing those that were used by a Mr. Blyth for an osteological lecture delivered before the Zoological Society of London in 1837 (" Proceedings, Zoological Society," 1837, p. 122). In the *Ibis*, 1860, p. 397, Professor A. Newton refers to these bones, and supposes that they were extracted from a skin. He lately informed Professor Wh. Blasius that Mr. Blyth afterwards gave him confirmation of the conjecture. It is not known what afterwards became of these bones, and with as little certainty can it be stated out of what skin they proceeded. In a letter to Professor Wh. Blasius, Professor A. Newton says—" I remember that many years ago I asked Blyth about the bones which he mentioned in 1837, but he could not remember out of what specimen they were procured, but it was probably one that was stuffed by Bartlett, the present superintendent of the Zoological Garden, London, for Tucker, a London dealer, who on one occasion about that time had from eight to ten skins of this bird all at once ; all of course from Iceland. Tucker died many years ago, and I never knew him. Bartlett has no remembrance of the affair, but he and Blyth were very intimate at that time."

The large quantity of remains of *Alca impennis* obtained on Funk Island during 1874 by Professor J. Milne, have been purchased by many museums as well as private collectors, and have proved useful in filling up in numerous instances a much felt want.

It appears that on the return of Professor Milne, he allowed the authorities of the British Museum and the Royal College of Surgeons, London, and Professor A. Newton of Cambridge, to make their selections, and afterwards handed over the greater part of his find to Mr. Edward Gerrard, junr., dealer in Zoological wares, who out of them put together the three skeletons already mentioned, and then sold most of the numerous remaining bones to his own customers and to other dealers who resold them. In our lists we have occasion to refer to these remains repeatedly. After deducting the three skeletons constructed by Mr. Gerrard and the one skeleton in the British Museum, South Kensington, we think the probable total we mention at page 86 is as near the correct number of Garefowl represented by the remains, as it is now possible to ascertain.

What may be the worth of a skeleton, or even detached bones, of *Alca impennis* is exceedingly doubtful, as almost no information on this point, as far as prices given in the past, is obtainable. The dealers, through whose agency most of the recent sales have been effected, are naturally reticent as to the prices paid them for remains by their clients.

In 1876 there was bought from Mr. E. Gerrard, junr., for the Museum of Science and Art, Edinburgh, an imperfect crania, two humeri, and several other bones from Funk Island, at a cost of 25s.

Physiological Preparations.

During 1883 we were informed by Professor Japetus Steenstrup of Copenhagen that he intended to have the viscera of the two Great Auks in the Museum figured. As we thought it might be possible to arrange to give prints of these figures in this work, we wrote the learned Professor, and he kindly replied as follows on the 25th August 1884:—"If the figures of the viscera of the Great Auks in the Museum had been ready I certainly would have sent you them. We have obtained specimens of *Alca torda* of both sexes from the Faröe Islands, killed on the same days of the year as our *Alca impennis* ♀ and ♂, and preserved in spirit. These reached Copenhagen some weeks ago. As soon as time allows, preparations similar to the preparations of *Alca impennis* are to be made, and then both sets of preparations are to be figured for comparison. Without suits the figures of the viscera of *Alca impennis* would be of no use at all." In the *Ibis* for 1861, page 390, Professor Newton refers as follows to the body of a Great Auk that was preserved in spirit:—"In August 1840 or 1841, three skins, as many eggs, and the body of a bird in spirit was bought of Factor Chr. Thaáe, now (1861) living at Copenhagen, by Herr S. Jacobsen, who told us that he parted with them either to Herr Selning, a naturalist at Hamburg, or to Mr. Jamrach, the well-known dealer." These remains had been evidently obtained on Eldey. What became of the bird in spirit appears to be unknown, and it would be most interesting if it could be traced, as physiological preparations of the Great Auk are so rare.

EGGS.

The value of the eggs is better known, as on one or two occasions they have been sold by auction, and the rapid increase in value is easily seen by an examination of the prices paid.

The two eggs now in Philadelphia were both bought from dealers in Paris by their previous possessor, M. O. Des Murs, at remarkably low prices compared to what are now given. The first was bought 3d June 1830 for 5 francs, about 4s. 1d. The second was purchased on the 10th May 1833 for 3 francs, about 2s. 5d. One of the eggs now in Cambridge is believed to have been bought in 1832 for £2. The specimen now at Breslau was sold at Leipzig in 1835 for 7 thalers = £1, 1s. It is said to have been purchased by the present holder in 1870 for 200 thalers = £30. Another egg, now at Scarborough, was also bought at Leipzig in 1835 for 7 thalers = £1, 1s., along with some other eggs. This Great Auk's egg was sold alone in 1857 for 50 thalers = £7, 10s., but what the present holder paid for it we do not know. Many years ago, probably about 1840, the egg recently brought to light at Poltalloch was bought by Mr. Malcolm from the elder Leadbetter for £1, which the purchaser thought at that time a high price.[11] The egg now at Hitchin was bought by Reid of Doncaster, in May 1841, for £2, 6s., from Friedrich Schultz of Dresden. In connection with this transaction, the following letter from Mr. Reid to Mr. R. Champley of Scarborough is interesting :—

"8 SPRING GARDENS, DONCASTER,
26th July 1860.

"DEAR SIR,—I never had but one egg of *Alca impennis* in my possession. I had the above from Mr. Friedrich Schultz of Dresden in Saxony. I sold it to Mr. Tuke of York, now banker in London (now Hitchin). I received it in 1841, May 23d. I sold it for £2, 6s. to that gentleman; it is now worth £28 or £30. The year following the late

---

[11] Mr. Malcolm, in letter of 23d August 1884.

Mr. Robert Dunn of Hull offered me two fine eggs of that bird, and two skins of the same for £20. I did not purchase, and I do not know who got them. I believe he had them from Norway. He has two sons in Orkney, who collect specimens of natural history. These are all the particulars I can give.—Yours truly,   Hugh Reid."

On the 30th December 1859, Mr. W. R. Johnston of Hackney sent to Mr. R. Champley a copy of a letter from the late Sir William Milner, Bart., which, however, has no date. It is written from Nunappleton, Tadcaster. The following is what Sir William says :—"Whilst I was staying at Dusseldorf, in November 1847, I heard that there was a Great Auk's egg to be had at Perrot's, an out-of-the-way shop down by the Seine in Paris. As I was returning to England I stopped in Paris, found that the information was correct, and purchased the egg from Perrot for 200 francs (about £8, 3s. 4d.), on the 23d November 1847. It is now in my possession, and considered a very good specimen in good preservation." This egg is still in the collection at Nunappleton (see p. 88).

During May 1853 the late Lord Garvagh bought at a public sale two eggs of *Alca impennis* that belonged to Mr. T. H. Potts, now of Ohinitahi, New Zealand, for £30 and £29. One of these has since got broken, and is understood to be in the possession of the Dowager Lady Garvagh. The other is in the collection of the late Mr. G. D. Rowley, which now belongs to his son Mr. G. Fydell Rowley, Brighton. Writing to Mr. R. Champley, probably about 1860 (though the letter bears no date), Herr Friedrich Schultz, Dresden, says : " The egg that you (query *Dresden;* this note is in Mr. Champley's writing) possess out of Thienemann's collection is from Paris, and was found there in the year 1845 amongst a heap of rubbish. I saw the egg at Paris in 1846, and could not recognise it for dirt as an *impennis*, therefore did not buy it. After my return I informed Dr. Thienemann of it, who sent for it, cleaned it, and kept it in his collection. He paid 15 francs for it." In a postscript to the letter he says : " One egg in Thienemann's collection is valued at 100 dollars."

In a letter to Mr. R. Champley, dated from Paris, 21st March 1860, a writer whose name we have not been able to decipher, mentions as follows :—" The egg bought at Paris by the Abbé Vincelot of Angers has been purchased for the collection of M. Raoul de Baracé of the same town. (See p. 88.) I learned some days ago that an egg of the same bird exists at Paris in the collection of Mons. Cerveau,[12] head of the department of the Minister of Public Instruction, and another at Dieppe (see p. 89), in that of M. Hardy, a learned ornithologist."

---

[12] We are unaware what has become of this egg.

On the 14th April 1860, Mons. E. Fairmaire of Paris wrote Mr. R. Champley informing him that he had obtained drawings for him of the two eggs at that time in the possession of Count de Baracé, Angers. He adds, " The egg of *Alca impennis* with the black streaks at the thick end, was sold by me to M. de Baracé for the sum of 450 francs (about £18, 7s. 6d.) "

The following curious story, which is well known to ornithologists, is so remarkable that we repeat it, and give a copy of Mr. R. Champley's original note, dated 1st June 1860 : " Mr. Bond says to R. C., —— Yarrell told him that walking near a village near Boulogne he met a fishwoman having some guillemot's eggs. He asked her if she had any more ; she said she had at her house. He went, when he saw hanging over the chimney-piece four wild swans', with a Great Auk's hanging in the centre. She asked two francs each for them. He bought the Auk's and two swans'. She said her husband brought it from the fisheries. The Great Auk's egg sold at Stevens' sale to Mr. Gardiner for £21 ; sold again by him to Mr. Bond, 24 Cavendish Road, St. John's Wood, London, for £26. Copied by R. Champley at Mr. Bond's, by whom the history was told."

On the 11th July 1865 there was sold at Stevens' Sale Rooms, London,[13] four Great Auk eggs, that were part of the splendid set of ten eggs discovered in the Museum of the Royal College of Surgeons. The prices they fetched were £33, £31, 10s., and two £29 each. As it may be interesting for our readers to have some further information about these eggs, we may state, that in a footnote to page 483 of " The Garefowl and its Historians " (" Natural History Review," 1865), Professor A. Newton mentions that a few years prior to that time there was found in the Royal College of Surgeons, London, by the late curator, Mr. Stewart, a box with the words, " Penguin's eggs—Dr. Dick "—" when or how they came into the possession of the establishment there was no record."

The box contained ten matchless Great Auk eggs, which were recognised by Professor A. Newton, and from the name Penguin being on the box he supposes them to be of American origin. This collection appears to have been unique and unrivalled, and to all interested in such remains invaluable for comparison. The authorities of the Royal College were evidently unappreciative of them, for it is stated that they disposed of some without even taking casts or photographs. From a letter which Mr. J. C. Stevens, the auctioneer, wrote to Mr. R. Champley

---

[13] J. C. Stevens Sale Rooms, 38 King Street, Covent Garden.

of Scarborough, dated 14th July 1865, we get the following information—" Lot 140, sold for £29, to the Rev. G. W. Braikenridge. Lot 141, £33, Mr. G. D. Rowley. Lot 142, £31, 10s., Rev. H. Burney. Lot 143, £29, Mr. Crichton." From other sources of information we learn that these eggs are now in the following collections:—Lot 140, which was bought by the Rev. G. W. Braikenridge of Clevedon, Somerset, has been recently purchased, along with a collection of eggs from that deceased gentleman's sister, by Mr. Edward Bidwell, of Fonnereau House, Twickenham. Lot 141 is still in the late Mr. G. D. Rowley's collection at Brighton, which now belongs to his son, Mr. G. Fydell Rowley. Lot 142 is still in the possession of the Rev. Henry Burney, Wavendon Rectory, by Woburn, Bedfordshire. Lot 143, which was purchased by Mr. Crichton, is now in the possession of his brother-in-law, Lord Lilford. In addition to the above four eggs, of which we have given the sale prices, other three from the same collection were sold privately to Mr. R. Champley of Scarborough, through the agency of Professor Flower, so the Royal College is now only in possession of three eggs of the ten that belonged to it.

On the 27th April 1869 there was sold, at Stevens' Sale Rooms, the egg that belonged to Dr. Troughton, which was purchased by the late Lord Garvagh for £64. The egg is now at Brighton. After Lord Garvagh's death his executors sold by private bargain this egg, and also one of the eggs which his lordship had bought in 1853 from Mr. T. H. Potts, to the late Mr. G. D. Rowley, and at the same time offered the fragments of the other egg purchased in 1853; but Mr. Rowley did not buy the broken specimen, which is probably still in the possession of the Dowager Lady Garvagh. We have heard, but we cannot be certain of the truth of the statement, that this egg got broken through the carelessness of a footman, and that it was with the object of replacing its loss that Lord Garvagh, in 1869, purchased the egg that had belonged to Dr. Troughton.

With regard to how this egg came into the possession of Dr. Troughton, the following letter is interesting :—

"COVENTRY, *8th Feb.* 1861.

" DEAR SIR,—I send you the drawing I promised you a long long time ago. It is of the exact dimensions of the original egg of the Great Auk in my possession, which I purchased of Mr. Bartlett, ten years ago, I think, with the bird. The markings are pretty faithfully made, considering it was my only attempt at egg drawing. I hope it will reach you safely.—I am, dear sir, yours faithfully, NATH. TROUGHTON."

" R. CHAMPLEY, Esq., Scarborough."

In the Museum of Science and Art, Edinburgh, are preserved the two magnificent specimens of the eggs of *Alca impennis*, of which, through the facilities kindly afforded us by the authorities of the Museum, we are able to give coloured plates at page 108. Nothing is really known as to the origin of these eggs except what we mention at page 87; but it is probable that they originally came from Newfoundland, as on one of them (No. 1 on plate) is the word " G. Pingouin." The egg, No. 2 on plate, has been at some time or other suspended by a string, as Mr. John Gibson informs us that on one occasion he was examining it when he thought he observed something inside, and with a little care at last managed to pull out through one of the holes in the shell a piece of string, to which was attached transversely a small stick. Both these eggs are blown at the ends. " Professor A. Newton (in *Ibis*, 1861, p. 387) mentions that Mr. Scales saw several eggs of *Alca impennis* in the hands of Mons. Dufresne in Paris about 1816 or 1817, and that Mr. Scales got an egg from him reported to have come from the Orkney Islands, which, however, Professor Newton thinks extremely unlikely, and hints that possibly the eggs may have been obtained at the Geirfuglasker about 1819, when it was rumoured a French vessel had visited the skerry." It is not mentioned when Mr. Scales got the egg from Dufresne, but if it was in 1816 or 1817, this egg could not have been part of the spoils of an expedition to the skerry in 1819. It is quite possible that Dufresne may have obtained the eggs in his possession from various sources, but this appears to us unlikely, and the eggs now in Edinburgh are evidently of American origin. We have heard doubt expressed as to whether the eggs in the Edinburgh Museum ever belonged to the Dufresne collection; but we think there can be no reasonable doubt upon that point, as they have always been associated with it in the Museum, and there has never been, so far as we know, the slightest suggestion as to any other source from which they could come, and the foregoing statement makes it evident that Dufresne was possessed of some eggs of *Alca impennis*. These, in the natural course of events, would be sold along with the other objects that formed his natural history collection. It has been stated by Professor Wh. Blasius in " Zur Geschichte der Ueberreste von *Alca impennis* " (page 157), " that the eggs were in the Edinburgh Museum quite unknown for fifty years." That is, however, a mistake, as the present curator of the Museum of Science and Art, Mr. Alexander Galletly, informs us that they were quite well known to be eggs of the Great Auk by the late Mr. James Boyd Davies, when he had charge of the Natural History Collection, and also to other persons. At that time remains of

Alca impennis were not so valuable or so much prized, and probably it was not thought of importance to publish the fact of their existence so as to record them. When Major H. W. Fielden became aware of these eggs being in Edinburgh, he ascertained the particulars, which enabled him to publish a short account of them (*Ibis*, 1869, pp. 358–360).

Some doubt has existed as to how the Dufresne collection was acquired by the Edinburgh University; but the following communication, dated 29th November 1884, which we received from Professor W. Turner, who holds the chair of Anatomy, makes the transaction quite clear: "The eggs of *Alca impennis* in the Museum of Science and Art formed a part of the collection of M. Dufresne of the Jardin des Plantes, Paris, which was purchased in 1819 by members of the Senatus of the University. It was afterwards acquired by the Senatus as a body, and was transferred by them to the Science and Art Department in 1855. The Dufresne collection consisted of about 18,000 specimens, and contained 1600 birds, and 600 eggs of birds, and many of the specimens were of great value."

There is at Breslau in Germany an egg which belongs to Count Rödern, and which is referred to at page 25 of Appendix; but as there are different opinions about its past history, we may mention what Mr. R. Champley says on the subject. He writes us under the date 23d September 1884: "In reply to your letter respecting the Breslau egg, I find the following memorandum, dated January 1861, from Mr. R. Mechlenburg, Flensburg. Mechlenburg sent me a drawing of the egg formerly in his possession, which he obtained direct from Iceland in 1830. He sold it to a dealer in Breslau. I copied the drawing, which I now have, and it is endorsed with the above particulars. I believe in the 'Für Ornithologie' [14] of that year there is reference also to the same egg—this is twenty-four years ago. I don't know whether it has changed hands; this is the authority I have for the egg at Breslau." On the 25th September Mr. Champley writes us: "I purchased a bird and egg of Mechlenburg, and at the same time he sent me the drawing of an egg he had sold at Breslau, which I copied." In the same letter Mr. Champley informs us that he has photographs, engravings, drawings, sketches, &c., of, he thinks, forty-four eggs, but not the two in the Edinburgh Museum."

In 1865, Professor W. Preyer, of Jena, mentioned an egg that existed in the

---

[14] A German publication, "Journal für Ornithologie.'

Egg of the Great Auk or Garefowl
(Alca impennis Linn.)
Preserved in the Natural History Collection
Museum of Science and Art,
EDINBURGH.

EGG OF THE GREAT AUK OR GAREFOWL
(Alca impennis Linn.)
Preserved in the Natural History Collection
Museum of Science and Art,
EDINBURGH.

Manchester Museum, but it appears that this egg has now been lost, and also a stuffed skin of *Alca impennis* that was purchased along with the egg for £13. (See page 95.)

About the beginning of May 1880 a collection of birds' eggs, shells, and other natural history objects was advertised to take place at Dowell's Auction Rooms, Edinburgh, and among others who received a catalogue was Mr. Small, bird-stuffer, George Street, in that city. He visited the auction rooms previous to the sale and examined the collection, and was struck with the appearance of two eggs marked "Penguin." At the sale he bought the lot which comprised the small collection of eggs for £1, 12s.

The two eggs having been examined by experts, were satisfactorily proved to be those of the Garefowl, and were sent to Stevens's Auction Rooms, London, during the following July, when they were sold separately, fetching £100 and £107, 2s., the purchaser being Lord Lilford. It is believed that these specimens are of American origin, and that they were obtained by a Dr. Lister, a surgeon in the army, and presented by him to his brother, Mr. Andrew Lister, who had a natural history collection, and lived in Edinburgh. On his decease they became the property of another brother, Mr. John Lister, advocate, who sold them to Mr. John Murray, S.S.C., in whose possession they remained about twenty-five years, and at his death fell into the hands of Mr. W. C. Murray, W.S., by whose instructions they were sold at Messrs. Dowell's. The story of these two eggs, which probably had been in Edinburgh nearly sixty years previous to their re-identification, shows that we need not yet lose all hope of a few stray remains of the Garefowl being brought to light in the future.

During the autumn of 1883 Lord Lilford bought another egg from Mr. G. A. Frank, dealer in zoological wares, London, for a price stated to be not much under £140. This egg was obtained from the Natural History Museum, Lausanne (see Appendix, p. 28). In a letter, dated 6th February 1885, Mons. Victor Fatio of Geneva informs Mr. R. Champley that he intends shortly to write a paper regarding this egg. On the 1st March 1885 Mons. Victor Fatio again writes to the same correspondent, "The egg No. 2 has been disposed of by a great mistake to Mr. G. A. Frank, London, in exchange for a bad skin of a Gorilla and a few other remains of little value. They tell me that Frank has resold this interesting specimen at a very high price in London, which is very probable. The Curator of the Museum, who now comprehends the great mistake he committed, is full of regret that he had not studied the subject more carefully."

Lord Lilford has lately succeeded in purchasing an egg in Dorsetshire, which is stated to be unrecorded, and we hear his lordship intends to publish particulars. The following information regarding this specimen was communicated by Professor Wh. Blasius to the meeting of the Natural History Societies of Germany at Magdeburg on 22d September 1884. He said, "Professor A. Newton of Cambridge informed him, on 19th August 1884, that an unrecorded egg had been discovered in Dorsetshire, where it had been in the possession of a family for many years, and is believed to have come originally from Newfoundland. It is now in the collection of Lord Lilford, and Mr. G. A. Frank of London informed Professor Blasius that his lordship paid £50 for it." We may add, what probably Professor Blasius was unaware of, that this egg was offered in the first instance to the authorities at the British Museum, but as they did not wish it, Mr. R. Bowdler Sharpe brought it under the notice of Lord Lilford, who succeeded in acquiring it from a Mr. Hill, who was its owner.

At the Magdeburg meeting Professor Wh. Blasius also stated that two eggs said to have been in St. Petersburg are not now known there. The learned Professor narrated that during his visit to Russia last spring (1884) he made every inquiry for Alcine remains in Warsaw, Kieff, Charkoff, Moscow, and St. Petersburg, but found none with the exception of the skin in St. Petersburg, which he concludes is the only remains of *Alca impennis* in the great Russian empire.

As we have gone into considerable detail regarding the value of the eggs of the Great Auk, it may not be out of place that we should refer to another standard of value besides rarity, by which the great masses of the public who never heard of such a bird are likely to estimate their worth. Mr. R. Scot Skirving tells an amusing story that well illustrates this.

In 1880, when the eggs that were sold at Dowell's Rooms, and purchased by Mr. Small, were discovered to be those of *Alca impennis*, it caused some excitement among ornithologists, and was the subject of general conversation. Mr. Skirving happened to mention the discovery to a popular Edinburgh minister and afterwards to a well-known newspaper reporter, and as neither of them had ever heard of such a bird as the Great Auk, he explained to them as well as was possible in a few words what it was like, and what made it of special interest. Both gentlemen gave him the same look of pity, and curiously enough expressed themselves in exactly the same terms. They said, "But the eggs are of no use, they will never hatch."

Rumours regarding Remains.

When the announcement was made in the magazines and newspapers that two eggs of the Great Auk had been discovered in a small private collection of eggs purchased at a public sale in Edinburgh for a mere trifle, it made the numerous hoarders of natural history objects begin to examine the collections of birds' eggs they possessed, which in many instances had been long put aside and forgotten. When the high prices which the eggs fetched in London became known, there was a rush made to dealers in natural history wares asking for information, and remarkable stories were told of how Great Auk eggs had been given away by mistake through their value not being understood. Much labour was expended in tracing such gifts to their possessors, in the hope they might be induced to part with them, but in every instance that we know of, it turned out that the eggs in question belonged to some other bird of the Auk tribe. Since that time ornithologists have been kept continually on the *qui vive* by rumours of discoveries, most of which have proved fallacious. There may, however, be a basis for some reports that have got abroad, even though nothing authentic can be ascertained. We have heard that a skin and egg of *Alca impennis* exists in or near Edinburgh, but the result of inquiry leads us to believe that the report has been exaggerated, as it has passed from one person to another, and that the only ground for the rumour is as follows :—Some time ago there is said to have died an old gentleman, who left a most valuable collection of birds' eggs, which were obtained by him many years since, probably early in the present century. Some of the eggs are very rare, and rumour asserts there is among them one of the Great Auk. This collection has passed into the hands of a gentleman who we hear is supposed to be not fully aware of their value. The person who knows him will not divulge his name for some private reason, and as even this individual has not had an opportunity of examining the collection, there is no certainty that the egg in question really exists.

While mentioning these vague reports, it is as well to state that at the time of the discovery of the two eggs purchased in Edinburgh by Mr. Small, a rumour went abroad that they at one time belonged to the Dufresne collection that was bought by some members of the Senatus of the University, which those who narrate the story say contained four eggs of the Great Auk at the time it came to Edinburgh. It is almost certain that Dufresne once had a number of eggs of the Great Auk in his possession [15] at Paris, but there appears to be not the slightest

---

[15] *Idem*, p. 107.

proof that more than two of these reached Edinburgh. Besides, it is to be remembered that the history of the eggs bought by Mr. Small has been traced back
to about the time that the Dufresne collection found its way to the Modern
Athens.[16]   There is every reason to believe these eggs came direct from Newfoundland to Britain, so the rumour seems to be without any good foundation.

IMITATION REMAINS.

Collectors will do well to be on their guard against purchasing imitations of
Great Auk remains, which unprincipled persons have been known to prepare with
great skill, the counterfeit eggs being clever copies of those that are genuine.[17]
It is only right, however, that we should mention that several well-known collectors have either manufactured, or had prepared for their own gratification,
imitations of the skins or eggs of the Great Auk, which are less costly than the
veritable article ; and the eggs, being exact copies of specimens in various collections, have a certain scientific value.   A writer who refers to this subject, says,
" The best casts are those by Mr Hancock, who can produce a drawer full to all
appearance of veritable Great Auk's eggs : in reality they are all shams but one,
but the resemblance is so perfect, that without touching, it is almost impossible
to say which is the real egg.   It took Mr. Hancock sixteen days to colour his
plaster imitation of the egg that belonged to the late (thirteenth) Earl Derby, which
was so foul when he received it that it had to be washed." [18]   Mr. H. E. Dresser
says, " The eggs of the Great Auk measure about $4\frac{35}{40}$ inches by $2\frac{27}{40}$ inches."
However, it is well to bear in mind that the eggs vary in size and markings, as
may be seen by the figures we give at page 108, which are exactly the natural
size, and give a good idea of the usual markings.

" Mr. Masters, of Norwich, possesses an imitation Great Auk, which, I am
told, is very good ; it was made by his servant, Samuel Bligh—the bill is of
wood.   Mr. Proctor, of Durham University Museum, has also manufactured a
Great Auk quite recently.   The black parts are Brunnick's Guillemot, and the
breast is a Northern Diver's ; and this fictitious bird, now in my possession,
contains a few feathers of the real Great Auk in the region of the neck." [19]

---

[16] *Idem*, p. 108.
[17] " Proceedings of Royal Society, Edinburgh," 1879-80, p. 682.   Mr. R. Gray.
[18] " Birds of Europe."   Mr. H. E. Dresser, vol. viii. p. 566.
[19] " The Zoologist," April 1869, p. 1643 ; Mr. J. H. Gurney, F.Z.S.

In the Darmstadt Museum there is an imitation specimen, the only part that is genuine being the head; but it must be a very good representation of nature, as Baron Edmond de Selys-Longchamps, Liege, writes concerning it in a " Note sur un Voyage Scientifique " in the " Comptes rendus des Séances de la Société Entomologique de Belgique, 1876," 7th October, p. lxx. : " A splendid *Alca impennis*, with the secondary wings well fringed with white, as in one of the specimens in the British Museum." The baron evidently must have seen it at a distance through the glass of the case, but still, it shows that this imitation specimen must be a very clever production to deceive such an authority on Alcine remains.[20]

As it was desirable, if possible, to obtain further information as to the materials of which this imitation specimen was constructed, we wrote to Professor G. von Koch, and also to Professor W. Blasius, and apparently the latter had also written to Professor G. von Koch, for, on the 3d September 1884, he wrote informing Professor W. Blasius as follows : " There remains part of the skull and one half of the bill, which are genuine. The skin, which is entirely imitation, is constructed with feathers of *Alca torda, Colymbus*, &c. Professor Koch opened up the head and some of the parts of the (Hornscheide) hornsheath, and found the bones filled up with agave pulp mixed with wax, with which, also, the whole bill was covered. The skull probably proceeded from the old cabinet of Natural History, where it had once belonged to a stuffed *Alca impennis*. Kaup caused the genuine skull to be covered over with wax, and then stuck over with sham feathers, and proceeded in this way until the complete bird was formed."

On the 31st December 1884 Professor G. von Koch writes us as follows : " The specimen of the *Alca impennis* in the Museum has been opened up by my directions, and a perfect skull has been found in it. The rest is badly put together out of pieces of skins of *Alca torda, Colymbus glacialis*, &c." [21]

ILLUSTRATIONS OF REMAINS.

The illustrations we have been able to give, we hope may enable ornithologists to avoid imposition. One of each of the different bones of the Great Auk found in Scotland and England have been figured, and in several instances the

---

[20] For remarks on the structure of the shell of the egg of the Great Auk, see Appendix VII., p. 40.

[21] For remarks on the possibility of making porcelain imitations of Great Auk eggs, see Appendix VIII., p. 40.

same bones of the bird, but from different localities, are given.   We regret the number of bones of the bird is so small, and we can hold out little hope that many of those awanting to form a complete skeleton will ever be found in Britain, as it will be noticed that among the few bones discovered there are several duplicates; and also, that those which have been brought to light are the bones in the body of the bird most likely to resist the ravages of time.   The probability seems to be that we have as yet only come across the last traces of a bird which was abundantly represented by its remains at the time such places as the shell-mound of Caisteal-nan-Gillean were formed, and we may reasonably conclude that if the examination of similar places is delayed much longer, archæologists will find no traces of this extinct bird to reward their labours.

CHAPTER XII.

USES TO WHICH THE GREAT AUK WAS PUT BY MAN.

THAT the Great Auk was good to eat there is abundant evidence, and that its savouriness as food has been known from early times is shown by the remains found in ancient kitchen-middens in Europe and America. The historical records regarding the voyages to North American waters in later times, give corroborative testimony, until we find that from the value put upon the bird as an article of food it was extirpated in that region of the world.

The French sailors, who hailed principally from Havre de Grâce, and who visited Newfoundland to fish on the Banks, depended greatly upon the Penguin or Great Auk for a supply of food.

A Mr. Anthonie Parkhurst writes a letter from Bristow, dated 15th November 1578, to Mr. Richard Hakluyt, of the Inner Temple,[1] in which he says: "*The Frenchmen that fish neere the grand baie doe bring small store of flesh with them, but victuall themselves always with these birdes.*" There is no reason to suppose that they were able to keep the bodies in a fresh state for any length of time, and from the notices we find in various works it would seem that the usual and generally adopted method was to salt them. This salting was a simple process, if we take the following statement made by Mr. Edward Haies,[2] as correct: "*It is stated the Frenchmen barrell them up with salt.*" This mode of curing was evidently continued as long as any Garefowls could be obtained, and was practised in Europe as well as America.

A Mr. George Cartwright, who writes in his diary[3] about Funk Island and the Penguins, under the date Tuesday, 5th July 1785, says—"The birds which the people bring from thence they salt and eat in lieu of salted pork." It was

---

[1] "Hakluyt's Voyages," vol. iii., pp. 172, 173. London, 1600.

[2] "Hakluyt's Voyages," vol. iii., p. 191. "Report of a voyage and the success thereof, attempted in the yeere of our Lord 1583, by Sir Humphrey Gilbert, knight, &c.," by Mr. Edward Haies, gentleman.

[3] "Journal of Transactions and Events during a residence of nearly sixteen years on the Coast of Labrador," by George Cartwright, vol. iii., p. 55.

not long after this time that the Garefowl became exceedingly scarce in the neighbourhood of Newfoundland; and no wonder, for the writer we have just referred to, on the same page from which we quote, states that "the poor inhabitants of Fogo Island make voyages there to load with birds and eggs. When the water is smooth they make their shallop fast to the shore, lay their gangboards from the gunwale of the boat to the rocks, and then drive as many Penguins on board as she will hold, for the wings of these birds being remarkably short they cannot fly. But it has been customary of late years for several crews of men to live all summer on that island, for the sole purpose of killing birds for the sake of their feathers : the destruction which they have made is incredible. If a stop is not soon put to that practice the whole breed will be diminished to almost nothing, particularly the Penguins, for this is now the only island they have left to breed upon." As the result has proved, what Mr. George Cartwright foresaw in 1785 has unfortunately been accomplished more thoroughly than probably he thought possible.

We have spoken of the practice of salting the Great Auk being practised in Europe, and we find it referred to in connection with the descent made upon the Geirfuglasker off Reykjanes, Iceland, by the crew of the schooner *Färöe* during July 1813, and it is stated that when they reached Reykjavik they had twenty-four Garefowls on board, besides numbers that had been salted.[4]

It is, however, to the American locality we have to go for most of our information regarding the uses to which the Garefowl was put, as the birds in early times occurred there in such numbers.

In the year 1540 one of the early voyagers mentions the loading of his two vessels with dead Penguins in less than half-an-hour, and states that besides what were eaten fresh there were four or five tons of them to put in salt. We can easily imagine that these early voyagers would use as many of the birds as possible in a fresh state, as these would afford them a pleasant and healthful change of diet after the salted food they would require to live on when crossing the Atlantic, which took much longer then than it does now.

We must remember, too, that the ships were sometimes for long periods fishing on the banks to get a cargo, from which circumstance fresh food became almost a necessity.

It appears that there were different ways of preparing the bodies of the birds

---

[4] Mr. J. Wolley's "Researches," *Ibis,* vol. iii., 1861, pp. 384-386.

either for eating fresh or salting; but this probably was not so much from any difference in the flavour given to the carcass by the mode of getting rid of its outer covering, as from the feathers in the early times being of almost no marketable value, while in later times they became an important factor in the traffic in its remains. As we have not met with any notice by the early writers that refers specially to this point, we have to draw inferences from some statements they make when referring to the slaughters of the Penguins.

When Mr. Robert Hore and other gentlemen visited the Penguin Island (situated off the southern coast of Newfoundland) in the year 1536, they captured many Penguins (see p. 6), and he says :—" *The foules they flead, and their skinnes were very like hony combes full of holes.*" He does not tell us how they *flead* the Penguins, but as he does not refer to the feathers, we think we may reasonably conclude that the skin and feathers were cast aside as useless, for if it had been otherwise, this writer, who gives minute particulars of what he and his friends saw and did, would have been almost certain to have mentioned it. In later times, when the feathers became an article of marketable value, the Penguins were in greater request than ever; and it seems that the feathers were at the period referred to the most valuable portion of its body, though the carcasses were also in demand as food. The process by which the feathers were detached from the skin was by the parboiling of the dead birds, and as there was no wood on Funk Island with which to feed the fires required to boil the water, the Penguin hunters used the fat bodies of the birds as fuel, and in this way consumed many carcasses. (See p. 30.)

We are told by perhaps rather a doubtful authority, Audubon, that some fishermen he met in Labrador told him that great numbers of the young of the Penguin were used for bait (see p. 7); and he says this information, along with the statement that the bird still bred (in 1838) on a low rocky island to the south-east of Newfoundland, was corroborated by several individuals in that country. From what is now known, it seems almost certain that Audubon was misinformed. Another author, Sir Richard Bonnycastle, writing in 1842,[5] refers to a trade in the eggs and skin; he says, " The large Auk or Penguin (*Alca impennis*, L.), which not fifty years ago was a sure sea-mark on the edge of and inside the Banks, has totally disappeared, from the ruthless trade in its eggs and skin."

Mr. Robert Gray, F R S E , mentions that in a letter received from Dr.

---

[1] " Newfoundland in 1842," vol. i. p. 232.

William Anderson, Heart's Content, that gentleman states he was informed by Joseph Bartlett, "that he had often heard his father, who died in 1871 at the age of seventy, speak of the Pinwing, and that crews occasionally got on the Funks, built enclosures, lit fires, and burnt the birds to death for pure mischief." Several other aged masters of fishing vessels, who have been spoken to by Dr. Anderson, recollect perfectly hearing their fathers refer to both birds and eggs which they had taken; and Mr. Smith especially referred to the eggs being of one pint capacity, and the feathers of the bird being of considerable sharpness, readily pricking the skin and causing festering. None of the aged people, however, examined by Dr. Anderson, seemed to be able to fix a precise date for the Penguins' disappearance from the Newfoundland habitats."[6] Writing of Iceland, Eggert Olafsson says, "The bird is very fat, has flabby flesh, and makes right good eating." The same author mentions "that the gullet and stomach of the Great Auk were turned to account by the fishermen, who prepared them as bladders, filled them with air, and used them as floats."[7]

The eggs of this bird appear to have been in great request in this locality, for an author who wrote about the middle of last century mentions as follows, regarding the bird rocks or Geirfuglasker off Cape Reikenes (*i.e.* Reykjanes):— "The inhabitants at a certain season go to these islands, though the expedition is very dangerous, to seek after the eggs of this bird, of which they bring home a cargo in a boat big enough for eight men to row."[8]

However curious the foregoing details regarding the uses to which the body of the Great Auk was put may be, the information we have to give regarding St. Kilda is more curious still. On such an isolated island as Hirta, better known as St. Kilda, where mankind must utilise all the advantages that Nature gives them to the fullest extent if they are to keep bare life within them, it would be astonishing if the islanders had not found out some curious use for some portion of the Great Auk's

---

[6] "Paper on two unrecorded Eggs of the Great Auk," by Mr. Robert Gray, F.R.S.E. "Proceedings Royal Society of Edinburgh," Session 1879-80, p. 678.

[7] "Reise igiennem Island," Travels in (or throughout) Iceland. Eggert Olafsson and Bjarne Povelsen, published at Soröe, Denmark, 1772.

[8] "Natural History of Iceland, &c," by N. Horrebow, London, 1758. With regard to what this author says, we may mention that owing to the violent Röst that runs between the mainland and these islands it would be necessary to visit the Geirfuglasker in a large boat; but from what is known of the numbers of the Garefowl that existed at this place about the middle of last century the cargo of eggs to which he refers must have been a very small one, unless we suppose there is a mistake in the English translation of his work (which is not to be depended upon). It is quite probable, if Horrebow means a full cargo of sea-birds eggs, that his statement is correct. (See p. 19.)

body. Besides using its flesh and egg they appear to have utilised its stomach, which, from being more capacious than that of any other fowl frequenting the island, was most suitable for their purpose. We feel sure it would puzzle most of our readers what that purpose could be; but we shall give, in his own words, the narration of the author to whom we are indebted for the information :— [9]

"*Hirta.*—The island of Hirta of all the isles about Scotland lyeth farthest out into the sea, is very mountainous, and not accessible but by climbing. It is incredible what number of fowls frequent the rocks there; so far as we can see the sea is covered with them, and when they rise they darken the sky, they are so numerous; they are ordinarily catched this way: a man lies upon his back with a long pole in his hand, and knocketh them down as they fly over him. There be many sorts of these fowls; some of them of strange shapes, among which there is one they call the Gare-fowl, which is bigger than any goose, and hath eggs as big almost as those of the ostrich. Among the other commodities they export out of the island this is none of the meanest.

"They take *the fat* of these fowls that frequent the island and stuff the stomach of this fowl with it, which they preserve by hanging it near the chimney, where it is dried with the smoke, and they sell it to their neighbours on the Continent as a remedy they use for aches and pains." [10]

The worthy knight who gave this narration to the author we have quoted evidently had vague ideas of the size of the egg of the Great Auk; but our author appears to have been aware of this, for in a later work [11] he places the Great Auk among birds of which he desires a more accurate description. We give the writer's own words—

Caput VII.

"De avibus quibusdam apud nos quae incertæ classes sunt, quarum proinde descriptiones accuratus desidero.

"Avis Gare dicta, Corvo Marino Similis, ovo maximo."

With regard to the other matters, however, he seems to have been pretty correct; and we get some further information about the use of the fat or giben from Martin,[12] which is interesting. He tells us that, "This *giben* is by daily experience found to be a sovereign remedy for the healing of green wounds," &c.

---

[9] "An Account of Hirta and Rona given to Sir Robert Sibbald by the Lord Register, Sir George M'Kenzie of Tarbat." We quote from John Pinkerton's "General Collections of Voyages and Travels," vol. iii. 4to., London, 1809.

[10] Sibbald, MSS., 33, 3, 2. Advocates' Library, Edinburgh.

[11] "De Animalibus Scotiæ." Edinburgh, 1684, p. 22.

[12] "A Voyage to St. Kilda," by M. Martin, 1698.

&c. " They boil the sea plants, *dulse* and *slake*, melting the *giben* upon them instead of butter. . . . They use this giben with their fish, and it is become the common vehicle that conveys all their food down their throats."

It appears that the giben was used as a universal medicine, and that it was composed of only the fat of their sea-fowls, which was stuffed into the stomach of a bird. In olden time when the Garefowl were plentiful it was that internal organ of these birds which was generally used, but in later times it became customary to use the same organ of the Solan Goose or Gannet, which remains the practice until now. The St. Kildeans find a use for almost every part of the birds they kill, and, as mentioned (at page 76), they were in the habit of employing, along with the remains of other sea-fowl, the bones of the Great Auk for manuring the portion of the island that they cultivate.

The time has passed for ever when the Great Auk or its remains can form an important item in the trade of nations; but the skins, bones, and eggs of this bird, which have realised large prices within recent years, will make for it a greater celebrity in the future, from the immense value that attaches to them; and weight for weight they will exceed by many times the worth of even gold. The craze of private individuals to hoard up objects and remains which can only be accessible to students in the public museums of our great cities, is to be sincerely regretted; and it is to be hoped that the possessors of the remains of the Great Auk, seeing such remains are so scarce, will show their public spirit by becoming benefactors, and placing them where they can be seen and appreciated by the poor naturalist or archæologist, as well as the rich commoner or noble lord.[13]

---

[13] In those pages we several times have the pleasure of mentioning donations of Alcine remains to our museums, but we may well call special attention to the great generosity displayed by John Hancock, Esq., to the Museum Newcastle-on-Tyne. His good example is worthy of being followed.

CHAPTER XIII.

SOME NAMES BY WHICH THE GREAT AUK HAS BEEN KNOWN.

IN almost every country inhabited by the Great Auk it appears to have had different names, or a different mode of pronouncing and spelling the names was adopted, and in several instances it had more than one name in the same locality. This has rendered it somewhat difficult for writers who have written about the bird to note all its occurrences, as they have not always been able to recognise it under strange names with which they were unfamiliar. It seems generally to have received its appellation from some peculiarity, or habit of life, which became the distinguishing feature, by which it was known to the human inhabitants of the lands the bird visited; and in giving it a name they seem to have sought to use expressions that referred to this distinguishing peculiarity or habit. For instance, the Greenlander called it Esarokitsok,[1] or the Little Wing, from the smallness of its wings and its being unable to fly. It is rather remarkable that though the name Geirfugl is of Scandinavian origin, it does not appear that the Norwegians ever had a name for *Alca impennis* that is attested on undoubted authority. The name *Geirfugla* occurs in "The Old Laws of Norway,"[2] but clearly refers to the *Geirfalkar* (*Falco gyrfalco*, Linn.), the Gyr-falcon.[3]

In answer to our inquiries, Herr A. Lorange, of Bergen, kindly informs us "that the different records of the occurrence of the Geirfugl at the Norwegian coasts during the last two centuries are not to be relied upon, and that not even in historic times has the Geirfugl been seen there by any naturalist, or mentioned in any authentic account." This is the statement of a competent authority; and we might pass from further consideration of the Norwegian names for *Alca im-*

---

[1] "Fauna Groenlandica," Otho Fabricius "Hafniæ et Leipsæ," 1780, p. 82. Crantz's Greenl. vol. i. p. 82.

[2] "Old Laws of Norway," edited by R. Keyser and P. A. Munch. Christiania, 1846, vol. ii. p. 471 (various reading).

[3] In a letter, dated 22d March 1885, Professor Steenstrup expresses the opinion that *Geir* in this connection with *Geirfalkar* is not of Scandinavian origin, but is derived from the Latin word *gyrus*, a circle, and is applied to birds that circle in the air.

pennis, if it were not that several names said to have been used in Norway are mentioned by different writers, and it is necessary to refer to their statements. H. Ström [4] says,—"Anglemager is the name given to a black and white sea-bird, in form *resembling the Alca*, but twice as large, and with a longer beak. It is distinguished by a white spot near each eye and by its very short wings, so that it should certainly be called Pingwin, or Anser Magellanicus authorum. I do not remember to have seen this bird cited by Norwegian authors, with the exception of Lucas Debes, who calls it Pingwin or Goifugl, and says that it is rare at the Faröe Islands. (Professor Steenstrup says, 'Here begins confusion with *Harelda glacialis*, the name of which is *Anglemaker*.') On the other hand, it is tolerably common with us. It appears in the bays as well as on the high sea in large numbers at the beginning of the spring fishing, crying continually, 'aangla,' as if to tell the fishers to get ready their *angler* (hooks), and that is why our fishers have called it Anglemager." M. Victor Fatio [5] says, "According to Ström, it (*Alca impennis*) was called Anglemager in the neighbourhood of Sôndmöre; but I ought to remark that in this transference of the name from the *Alca torda* (Razorbill) (*sic*, should be Havetlen or *Harelda glacialis*) to the *Alca impennis* (Great Auk), there is without doubt more than one mistake."

Professor A. Newton translates *Anglemager*, hook-maker (literally Angle-maker), but expresses doubt as to its being a name for the *Alca impennis*, and is inclined to suppose it to have been applied to the *Harelda glacialis* (Longtailed Duck) of modern naturalists. [6]

Brunnich [7] gives the name *Brillefugl* for the Great Auk, which means the spectacle bird, or the bird with the glasses, from the white patches on each side of the head in front of the eyes. This name was never known in Norway, and appears to have been invented by the Danish naturalist himself. Herr Preyer [8] says, "Brillefugl is a Danish name for the Great Auk." This is also a mistake, unless he merely means it was a name used by Brunnich.

Pennant [9] mentions *Fiært* as a Norwegian name for *Alca impennis* upon the authority of a Dr. Œdman, but evidently this gentleman has been under a misapprehension regarding the use of the name. The nearest approach to the word

---

[4] H. Ström, "Physical and Economical Description of Sôndmöre (near Aalesund, west coast of Norway), in the Department of Bergen, Norway," printed at Soroe in Denmark, 1762, vol. i. p. 221.

[5] "Bulletin de la Societe Ornithologique Suisse," tome ii. 1er part, p. 15 (footnote).

[6] "Natural History Review," 1865, pp. 469, 470.

[7] "Arct. Zool." vol. ii. p. 220.

[8] "Ueber Plautus Impennis," p. 17.

[9] Pennant's "British Zoology," vol. ii. p. 146.

that we have been able to find in either Icelandic, Norse, or Danish diction-
aries, are the Danish words Fjært, *peditum*, and fjærte, *pedere;* but some writers
have supposed that the name has been derived from *Fiante*, meaning simpleton,
sot, silly man. In the course of our search for further information regarding the
use of *Fiært*, we discovered in an old Norse dictionary (in which, it is stated,
are " shortly set forth various Norse glosses of daily phrases, the wonderful native
names of fishes, birds, and beasts, along with various proverbs," &c.) the word
Fiærskiit, or in modern spelling *Fjærskidt*, which is said to be the name given to
some small grey birds that run forward with the ebb in the expectation of feed-
ing themselves on the sea-weed that lies on the rocks. They obtain their name
from their dirtying of the beach.

As we were anxious to get all possible information upon the subject, we
communicated with Herr A. Lorange of Bergen, who, under the date of 12th
March 1884, kindly informed us, " that Dr. Œdman was probably Johan Œdman,
author of ' Chorographia Balmensis,' published at Stockholm in 1746, who was
not a critical man. He was a Swede, and not a Norwegian. His statement that
Fiært at any time has been a Norwegian name for the *Geirfugl* is a complete
misunderstanding. From Ström [10] we learn that the name Fjært, or Baare=
Fjært, on the coast *sometimes* is given to the *Alca alle*, Linn., the *Mergulus alle*,
Ray (Little Auk). This designation is not now known even in Sôndmöre. In
Pontoppidan [11] the same bird is called *Boefjær*. ' *Fjære* '—that part of a shore
where the tide water is going up. *Fiante* is a misreading. *Fjærskidt*, mentioned
by you from an old dictionary, is the fisherman's expression for the *Tringa maritinia*
(Purple Sandpiper). This dictionary of Christen Jensôn, priest at Askevold near
Bergen, is a *liber rarissimus*, and was printed (by Hahn) at Copenhagen in 1646."

The " Garefowl," or " Gairfowl," is the name by which the natives of the
Western Isles of Scotland must have known it for many centuries, and is the
name used by Professor Newton in his invaluable papers regarding the history of
the bird. We think there can be no doubt that this name is of Scandinavian origin,
and that it is part of the rich legacy of names of places, and things animate and
inanimate, left by the Scandinavian invaders of Western Scotland, who, under
the names of Danar or Danes, Dubh Gaills, or Black-haired Gaills, Finn Gaills,
or White-haired Gaills—the latter supposed to be Norwegians, and, perhaps, also

---

[10] " Physical and Economical Description of Sôndmöre " (near Aalesund, west coast of Norway), printed
at Soroe in Denmark, 1762.
[11] " Natural History of Norway." Copenhagen, 1752.

Swedes—began their depredations about the year 794 A.D.; and, after long exercising by their raids a terrorism over the Celtic population of the Isles, and especially over the inhabitants of the ecclesiastical settlements (these being the chief objects of their cupidity), ultimately settled down into the possession of the Hebrides as conquerors, occupying that position until, in July 1266, these islands once more returned to the Scottish rule. Though it is almost certain that this name came to us from the Scandinavians, it is not quite so easily determined from what particular habit or appearance of the bird they gave it the name; and on this point there is a difference of opinion, though all seem to be at one in the belief that such variations as Goirfugel,[12] Garfogel,[13] Goifugl,[14] Avis Garfahl,[15] Gaarfuglur,[16] Gyr-v-Geyrfugl,[17] Gorfuglir,[18] Garfowl,[19] Geyer-fogel,[20] Gairfowl,[21] Garefowl,[22] Avis-Gare,[23] Gare-Fowle,[24] Goifugel,[25] Geirfugl,[26] Geir,[27] Gar-fowl,[28] Goirfugl and Gaarfugl,[29] are all derived from one common origin.

The late Dr. John Alexander Smith, in his first paper on the Great Auk,[30] states that Geyrfugl is the name by which the bird is known in Iceland. He says, "*Geyr* is the Icelandic for a spear," and goes on to state that the terms *Geyr-fugl*[31] may mean the bird with the spear-like beak, or it may refer to the

---

[12] *Goirfugel.* "Exoticorum Decem Libri," Carolus Clusius. Leyden, 1605, p. 367.

[13] *Garfogel.* "Debes Færoa Reserata," published in 1673.

[14] *Goifugl.* "Synopsis Methodica Avium et Piscium," John Ray, London, 1713.

[15] *Avis Garfahl.* "Barthol. Act," p. 91, referred to by Carolus Linnæus. "Fauna Suecica, Lugduni." Batavorum, 1746.

[16] *Gaarfuglur.* "Landt Beskrivelse, over Færoeerne," published 1800.

[17] *Gyr-v-Geyrfugl.* These names are given by Pennant as being used in Iceland: "British Zoology," vol. ii. p. 146.

[18] *Gorfuglir.* The name given to the Bird in the Färoe Isles ("Nat. Hist. Review," 1865, p. 475).

[19] *Garfowl.* Professor Steenstrup in litt. 8th April 1882.

[20] *Geyer-fogel.* "A Short American Tramp in the fall of 1864." Campbell. This writer probably spelt the name from memory, as we have not met with this mode of spelling in any other work we have consulted.

[21] *Gairfowl.* "Voyage to St. Kilda," M. Martin, Gentleman. 1698, p. 27.

[22] *Garefowl.* "A Description of St. Kilda," by the Rev. Alexander Buchan, minister there from 1708 to about 1730, published by his daughter, 1773. Most of the book is made up of extracts from Martin's "Voyage," 1698.

[23] *Avis-Gare.* "De Animalibus Scotiæ," Sir Robert Sibbald. Edinburgh, 1684, vol. ii. p. 22.

[24] *Gare-Fowle.* Sibbald, MSS., 33, 3, 2. Advocates' Library, Edinburgh.

[25] *Goifugel.* "Synopsis Methodica Avium et Piscium," by John Ray. London, 1713.

[26] *Geirfugl.* "Proceedings Royal Society, Edinburgh." 1879-80, p. 680.

[27] *Geir. Ibid.*

[28] *Gar-fowl.* "Transactions of Zoological Society." London, vol. v. pp. 317-335.

[29] *Goirfugl* and *Gaarfugl.* "Ueber Plautus impennis," von William Preyer, Heidelberg. 1862, p. 17.

[30] "Proceedings of the Scottish Society of Antiquaries," vol. xiii. p. 83.

[31] Herr A. Lorange, of Bergen, in a letter to us, dated 12th March 1884, says, "*Geier* in the old Northern (Icelandic) means = a pointed spear." In Anglo-Saxon the word *Gar* has the same meaning. We may mention that *Gai* was the Celtic name for a spear. That the upper mandible had some resemblance to a spear-head maybe gleaned from the curious mistake that is referred to at page 45.

extraordinary swiftness of the bird in the water being like the flight of a spear.

Professor Steenstrup agrees with Dr. Smith in thinking that the word *Geyr-fugl*, or more properly *Geirfugl*, means the bird with the spear-like beak, but thinks it improbable that the name refers to the extraordinary swiftness of the bird in the water, being like the flight of a spear. He believes the name is derived from its elongated spear-like or sword-like bill, in comparison with that of the Razorbill (*Alca torda*, Linn.) The learned Professor has been kind enough to furnish us with the following remarks : — " The word *Geir, Geyr,* or *Geyerr* in the name *Geirfugl* (*Alca impennis*, L.), is certainly of Icelandic or Scandinavian origin, and it signifies a spear or spear-like weapon or instrument. But the word *Geir* in Geirfugl when the name refers to Falco albicans, otherwise Falco gyrfalco, is of Latin origin, and derives from *gyrus, gyrare*, to make whirls or circular evolutions in the air. As a technical expression it has come from the old falconry-art, and has been adopted into " The Old Laws of Norway " (see p. 121). The name Geirfugl when applied to *Alca impennis*, L., is clearly Scandinavian (Icelandic). It means the Auk with the longer spear-like bill, just as you in English name the Little Auk, Razorbill (*Alca torda*, Linn.); and as I in Krisavik (South Iceland) in 1839 and 1840 always got the *Uria troile* (or hringvia), with the longer and more pointed bill, under the name *Geirnefia*, and the *Uria Brunnicher* with the shorter bill under the name *Stuttnefia*, meaning in English *Spearbill* and *Shortbill*. As a parallel I may remind you of the Icelandic and English names of *Esox belone* (*vulgaris*)—in Icelandic *Geir-sil ;* in English *Garfish, Garpike*, on account of the long and pointed snout."

In considering this question we must bear in mind who were the early settlers in Iceland, and endeavour to see if their previous history gives us any clue as to how this name for the Great Auk could have originated. As far back as the eighth century of the Christian era, there were monks of the Columban Church in Iceland, and it is believed they were the first settlers. An Irish monk named Dicuilius wrote a work in the year 825 A.D., in which he mentions that at least thirty years previous to that time he had seen and spoken with several monks who had visited an island they called Thile, which has been clearly proved to be Iceland from statements regarding the length of the days at different times of the year, and a calculation of the duration of the seasons, which accompany the story. Those monks had their settlements all the way north from Ireland through the

Western Isles of Scotland to Orkney, Shetland, Faröe, and Iceland.[32]   At this latter island they had a settlement in the West-Mann Isles off the south coast, which get their name from the anchorites coming from the West Land, or the country that borders the west of Europe, namely, Ireland.

On one of those islets the Great Auk bred, and although one of those stations where it has long been exterminated, we cannot doubt that at the time those islets were inhabited by the Christian monks it bred there in great numbers, and they must have been quite familiar with the bird.   Whatever was the name by which they knew it, we do not think it was the " *Geir-fugl*," as that appellation is evidently Scandinavian, while the name by which those monks called it would be Celtic in its origin.[33]   Those anchorites were not colonists in the ordinary sense of the word, for they were no lovers of women, and retired to those solitudes for the purposes of prayer and religious exercise, desiring to worship God in peace.

The next settlers in Iceland were men of a different stamp, who were heathens, but who came to the island to get rid of the controlling power of government as put forth by King Harold Fairhair in Norway, which made them to be the "king's men at all times."   They were no servile or savage race; they were freemen born and bred, brave warriors, and adventurous seamen, who had vested rights and world old laws that they would not allow to be interfered with. Rather than suffer kingly rule, they, family after family, left their Norwegian homes and sought a new settlement in Iceland.   The first settler who arrived from Norway in Iceland was Ingolf, who came in the year 874 A.D., and ere long he was succeeded by others, until most of the habitable parts of the island were allotted to various families.   It is to these people we are indebted for the name the Great Auk bears in Iceland, and they may have brought the name with them.   At the time in which they or their forefathers lived, the Great Auk was most likely met with occasionally off the Norwegian coast,[34] but at any

---

[32] "Saga of Burnt Njal," Dasent. Introduction, pp. vii. and viii. ; also "Dicuile Liber de Mensura Orbis Terrae. Ed. Valckenaer," Paris, 1807 ; and Maurer "Beiträge zur Rechtsgeschichte des Germanischen Nordens," i. 35.

[33] Mr. R. Gray, "Birds of the West of Scotland," p. 441, gives "An Gearbhul" as the Gaelic name for the Great Auk ; but we have been unable to discover the source of his information, and suspect it is merely a Celtic corruption of the Norse name. The Rev. John Lightfoot (afterwards Dr. Lightfoot), who was the friend of Pennant, published his "Flora Scotica" in 1777, and at the commencement of it gives a list of the fauna of Scotland along with the Gaelic names, when he could discover that any such names existed ; but though he gives the Gaelic name for the Razorbill, "*Coltraiche*," he mentions no Gaelic name for the Great Auk, and he had every opportunity of ascertaining if there was one as he travelled through the Hebrides.

[34] Professor Steenstrup, writing us on 22d March 1885, referring to the results of investigations as to the Great Auk having frequented the Norwegian coast, says, "We now know very well that during historic times

rate was an inhabitant of the fiords of Denmark. There can be little doubt it was those Norwegian Vikings who spread the name of Geir-fugl through all the lands they settled in or cónquered, which name, as we have shown, has now been varied in many ways as to spelling and pronunciation in different countries.

We think there can be no doubt that the Scandinavians gave the *Geirfugl* its name on account of its spear-like bill, as stated by Professor Steenstrup. It is generally supposed that the earliest mention of the name *Geirfugl* occurs in the Edda,[35] where it is understood to refer to the Great Auk, *Alca impennis*, Linn., though the following gleda (Falco milvus) makes it uncertain.[36]

The evidence we have just given seems so conclusive that it might be accepted as finally deciding the question, if it were not that some persons have supposed that *Geir* in *Geirfugl*, and the German word *Geier*, meaning *Vulture*, are one and the same as regards meaning. This misconception was probably caused in the first instance through an awkward mistake made by the translator of the English edition of Niels Horrebow's " Natural History of Iceland," which was published in London in 1758. He probably translated from the German edition, and mistook the word *Geir* for *Geier*, or its older form *Geyer*. This led him to refer to the Great Auk as the *Geir* or *Vulture*. " The Vulture Rocks, called also Bird Rocks, lie beyond Reikenes, in the south district, about six or eight leagues west of this place. On these cliffs and rocks are a great many Vultures, which besides harbour in other parts of the island." Horrebow devotes a considerable part of his work to calling in question statements made by Herr Johann Anderson in his work on Iceland,[37] and it seems probable that, in some instances at least, Anderson was rather credulous in receiving information from those who aided him during his inquiries in Iceland. Horrebow quotes Anderson to con-

---

the Geirfugl has not been seen at the Norwegian coast, and we also now are better acquainted with the conditions of the sea-shores and sea-bottoms in such countries where the bird lived in former times, and hence it is no longer difficult to understand why it did not live at the Norwegian coast. I am now of the opinion that the whole range or line of the Norwegian coast from north to south is environed with such deep water that the Geirfugl by diving could not get to the bottom of the sea, in order to catch its food there. Consequently I am not disposed to think that the inhabitants of Norway or the Norwegian Skers (*Islands*) did see more than a single or bewildered Geirfugl, and that exceedingly rarely, nor heard the name given to the Geirfugl (not in Norway) but in the Faröe, Western Islands, or Iceland."

[35] " Prose Edda." Edition of Copenhagen, 1848-1852, vol. ii. page 483.

[36] Professor Steenstrup, in a letter to us, dated 22d March 1885, says—" Although the word *Geir* is Scandinavian, it is not properly a Norse word in its connection with Geirfugl. That is to say, this name of the bird, *Alca impennis*, L., is certainly given in countries where the bird was, if not very common, at least not exceedingly rare."

[37] Herr Johann Anderson (sometime Burgomaster of Hamburg), " Nachrichten von Island, Grönland und der Strasse Davis," &c. Frankfurt u. Leipzig, 1747, p. 52.

tradict his statements, and the translator thus gives one of these quotations in English:—" The *Geir* or *Vulture* is not often seen in Iceland, except on a few cliffs to the west, and that the Icelanders, naturally superstitious, have a notion that when this bird appears it portends some extraordinary event. Of this he (Anderson) assures us, being told 'that (in 1729) the year before the late King Frederick IV. (*of Denmark*) died, there appeared many, and that none had been seen before for years.'" There are other passages in Horrebow's work that are mistranslated, the *Geirfugl* being called the *Vulture;* but we need not refer to these, as the quotations we have just given are sufficient to illustrate how the mistake in the name has occurred. Mr. Robert Gray [38] refers to the name *Vulture*, which he supposes Horrebow had given to the Geirfugl, being no doubt unaware that there was a mistake in the translation, and then goes on to remark—

" Whether this writer had traced any connection between the Iceland name *Geirfugl* and *Lammergeir*, or *geyer* (literally, 'Lamb Vulture') which is a connecting link between the eagle and the vulture, I am not prepared to say—the etymology of the name Garefowl being confessedly a difficult question. Professor Newton informs me that the obvious resemblance at first sight between *Geir* and the German *Geier* or *Geyer* (its older form) has struck several persons, but that he doubts if it is more than a coincidence." Professor Steenstrup, in a letter to us, dated 22d March 1885, remarks " that he cannot understand why the etymology of the name *Geirfugl* should be a difficult question, because it is clearly a Scandinavian (Icelandic) name, and means the Auk with the spear-like bill." As to the suggestion of *Geir* in *Geirfugl* as *Alca impennis*, meaning *Vulture*, " the quoted respectable Niels Horrebow " does not indicate any connection at all between *Geirfugl* and *Vulture*. Both the Danish original and German editions have *Geirfugleskjer* or *Fugleskjer*, and it is only by a horrible mistake that the English translator (I think from the German edition) has translated the name of the rocks as "the Vulture Rocks." Nor does the map of Iceland in Horrebow's work contain any hints as to *Vultures;* even in the German edition the map has " Geir-Vogel " oder " Vogel-Schar." Referring to the statement of Anderson quoted by Horrebow, " the *Geir* or *Vulture* is not often seen in Iceland, except on a few cliffs to the west," Professor Steenstrup says, " The German edition (and the Danish in the same way) has only ' Der *Geyr* Vogel wird gar selten gesehen und zwar allein an den unten her an der

---

[38] " Paper on two unrecorded Eggs of the Great Auk," R. Gray, Esq.   " Proceedings Royal Society, Edinburgh," 1879-80, p. 680.

Westseite liegenden Klippen,' (p. 203)." The resemblance at first sight between *Geir* in *Geirfugl* and the German word *Geier, Geyer,* &c., is not so obvious as supposed, the first never being used without having *Fugl* or *Fowl* behind it; the German word *Geyer* never being used with *Vogel* behind it, without changing the word's meaning.

As some persons who are quite agreed with us as to the name *Geirfugl* being Icelandic, are not quite so certain that the name was given to the Great Auk on account of its spear-like bill, and are inclined to leave the name an open question, we may state their views. They think it possible that perhaps the progenitors of the Icelanders brought the name with them from Scandinavia, and again their forefathers had obtained it in the south of Europe, and they suppose that *Geirr,* the Icelandic for a spear, and the German word *Geier,* a Vulture, may have been derived from the same root and have similar meanings, and that the early Scandinavian settlers gave the name to the *Geirfugl* thinking it somewhat resembled a *Vulture.* We think there can be little doubt that this view was strengthened by the supposition that the English translation of Horrebow's work was correct, and that the Great Auk had been known in Iceland by the name of "Vulture." Professor Steenstrup, writing us 22d March 1885, says—"This argument, I think, has really no basis at all, our Scandinavian progenitors in the East being supposed to have had a quite different name for *Vulture* than *Geyer;* and again, the word *Geyr* or *Geyer* in the signification of *Vulture,* is not of Scandinavian or Northern root at all. You may be sure the idea of a *Vulture,* or the tradition or reminiscence of a *Vulture,* or a Vulture-like shape and behaviour, has nothing to do with the old Scandinavian conception of their Geirfugl (*Alca impennis,* Linn.)." [39]

One writer, the Rev. Mr. Kenneth Macaulay,[40] who wrote a history of St. Kilda or Hirta, mentions the Garefowl. He says, "The men of Hirta call it the Garefowl, corruptly perhaps instead of Rare-fowl—a name probably given it by some one of those foreigners whom either choice or necessity draw into this region." It is evident from the foregoing that Mr. Macaulay did not know of the Norse origin of the name, but it is also evident that he recognised that it was not Celtic. He appears to have proceeded to account for the name on the

---

[39] In Cleasby and Vigfusson's Icelandic-English Lexicon, the following is given regarding the word *Geirr*: M. [Anglo-Saxon *Gar,* Heliand *Ger,* Old High German *Keir,* whence *Kesja,* q. v. cp.; also Latin Gaesum. A Teut.-Lat. word] a spear.

[40] "The History of St. Kilda," by the Rev. Mr. Kenneth Macaulay, minister of Ardnamurchan, missionary to the Island from the Society for Propagating Christian Knowledge. London, 1764, p. 156.

assumption that it must be of English origin, and probably was under the impression that the ears of the Celtic inhabitants of St. Kilda were not sufficiently acute to distinguish the difference of sound between Rare and Gare, in a language with which they had at best only a very slight acquaintance. However this may be, no one need doubt that the Rev. Mr. K. Macaulay made a mistake as to the derivation of the name. It is interesting to notice the close association in the popular mind of the Great Auk and the Razorbill, *Alca torda*, L., from the similar names given to both these birds in different countries, and to which we have more than once to refer. For instance,—*the Gurfel* is mentioned by M'Gillivray [41] as being an appellation by which the Razorbill was known, and Fleming [42] gives *the Garfil* as a Welsh name for the same bird.

Great Auk.—The name " Great Auk," by which this bird is likely to be known in all time coming to the great majority of persons, is, as far as we can discover, of comparatively recent origin. It does not appear to have been in use before the time when Linnæus gave the name *Alca impennis* to the Garefowl and *Alca torda* to the Razorbill, sometimes called the Auk; and it is probable the word " Great " in the name was originated to distinguish to the unscientific these two birds by the difference in their size. Pennant [43] gives the name " Great Auk," and quotes opposite it " Latham " [44] as his authority; on the same page he also gives the name *Alca major*, and refers the reader to " Brisson; " [45] but this only takes us back a little more than a century, while the names Garefowl and Penguin were given to this bird long before that time.

We are unable to say how this name came into such general use in preference to others; but perhaps it has been owing to the writers in most of the recent works on natural history, with a few notable exceptions, having adopted the appellation, and the bird having appeared also under this name in press notices and caricatures.

The King and Queen of the Auks.—The " King and Queen of the Auks " was the name given to the two last specimens of Garefowl killed in Orkney, and the inhabitants of Papa Westra [46] and adjoining islands gave them this appellation,

---

[41] " British Water Birds," vol. ii. p. 346.   [42] " British Animals," 1828, p. 130.
[43] Pennant. " British Zoology," vol. ii. p. 146.   [44] Lath. " Ind. Orn.," 791, id. syn. v. 311.
[45] Brisson. Av. vi. 85. Tab. 7 (1760).
[46] " Papa Westra " was the site of an ecclesiastical settlement of the Columban Church; the island gets its name from the anchorites, who were called " Papar " by the Northmen.

we suppose, to distinguish them from the *Alca torda*, L., or Razorbill, and perhaps also the *Arctica alle*, L., or Little Auk.

Penguin.—A name by which it was known in almost all the countries it inhabited during last century was "The Penguin," [47] or "le Grand Pingoin;" [48] and there can be little doubt that this name, which appears to be of Welsh origin, and which is now given to a class of birds inhabiting the Southern Hemisphere, was originally given to the Great Auk, and it was from some similarity in the habits and appearance of these birds when viewed from a distance, though they differ widely, that led the early mariners who visited the Southern Seas to confuse the *Spheniscomorphœ* with the *Alca impennis*, L., with which they were familiar in the North Atlantic. This confusion of the names led to no end of trouble among ornithologists, very few of whom fifty or sixty years ago really knew what were the points of difference between the species, and some were to be found ready to deny that such a bird as the Great Auk ever existed, but time and experience have changed all this, and *Alca impennis*, L., is now duly placed in its proper niche by the ornithological world.

Professor Steenstrup, in his admirable paper on the Great Auk, mentions that the name Penguin is of Welsh origin; and if so, seamen from that part of our islands may have been the first to give the birds a name in the American locality, unless we suppose that it had received this name when it was a frequent visitant to our own shores and before the early voyages to Newfoundland took place. In the course of our studies, we have met with the following interesting statement. It is found in the third volume of "Hakluyt's Voyages," and occurs in—

"A true Report of the late discoueries and possession taken in the right of the Crowne of England of the Newfound Lands, By that valiant and worthy Gentleman, Sir Humfrey Gilbert, Knight. Wherein is also briefly set downe her highnesse lawfull Title therevnto, and the great and manifold commodities that are likely to grow thereby, to the whole Realme in generall, and to the Aduenturers in particular; Together with the easinesse and shortnesse of the Voyage. Written by Sir George Peckham, Knight, the chief aduenturer and furtherer of Sir Humfrey Gilberts Voyage to Newfound Land."

---

[47] Idem, p. 5.

[48] Given by Pennant, "British Zoology," vol. ii. p. 146, who refers his reader to "Hist. d'Ois.," ix. 393. Pl. enl., 367.

The third chapter of this Report is headed as follows :—

" The third chapter doeth show the lawfull title which the Queenes most excellent Maiestie hath vnto those Countries, which through the ayde of Almighty God are meant to be inhabited.

" And it is very evident that the planting there shall in time right amply enlarge her Maiesties Territories and Dominions or (I might rather say), restore her to her Highnesse ancient right and interest in those Countries, into the which a noble and worthy personage, lineally descended from the blood-royall berne in Wales, named Madock ap Owen Gwyneth, departing from the coast of England, about the yeere of our Lord God 1170, arrived and there planted himself and his colonies and afterward returned himself into England, leaving certaine of his people there, as appeareth in an ancient Welsh Chronicle, where he then gave to certaine Islands, beastes and foules, sundry Welsh names, as the Island of Pengwin, which yet to this day beareth the same.   There is likewise a foule in the saide countreys called by the same name at this day, and is as much to say in English, as Whitehead, and in trueth the said foules have white heads.[49]   There is also in these Countreis a fruit called Gwyneths, which is likewise a Welsh word.   Moreover, there are diuers other Welsh wordes at this day (about 1583 A.D.) in use, as David Ingram aforesaid reporteth in his relations.   All which most strongly argueth the sayd Prince with his people to have inhabited there."

There appears on the margin of the page this note :—

" 1170. Owen Gwyneth was Prince of North Wales.   Nullum tempus occurrit Regi."[50]

For further particulars regarding Owen Gwyneth, see Appendix III. page 35.

Carolus Clusius figures the Penguin of the Southern Hemisphere, and says, " The name *Penguin* is derived from their excessive fatness (Latin, *pinguis*, fat) ; "[51] and if he is correct, the Great Auk or Penguin of the North Atlantic had doubtless received the name for the same reason long before.   Indeed, this is referred to by Mr. John Reinhold Forster, in his narration of the voyage of Mr. Robert Hore, who visited during the summer of 1536 an island on the southern coast of Newfoundland, named Penguin Island, and he says the island had got its name from a kind of sea-fowl, which the Spaniards and Portuguese called Penguins on account of their being *so very fat*.[52]   This derivation is one of the old-fashioned kind, and its absurdity does not need to be pointed out.

George Edward figures the Great Auk, and names it the Northern Penguin.[53]

---

[49] The Great Auk has a white patch on each side of the head in front of the eye.

[50] "Hakluyt's Voyages," London, 1600, vol. iii., pp. 165, 172, 173.

[51] "Exoticorum Decem Libri," Leyden, 1605, p. 101.

[52] "History of the Voyages and Discoveries made in the North, by John Reinhold Forster," I.U.D. London, 1786, p. 290.

[53] "Natural History of Birds," Part III., London, 1750, 4to, plate 47.

The specimen of the bird from which this figure was produced, he tells us, was caught by the crew of a Newfoundland fishing-vessel at the Banks a hundred leagues from the shore, where it was taken with their fish-baits. Edwards tells us he procured the bird from the master of the boat. Buffon [54] calls the bird *Le Grand Pingouin*, while Temminck [55] gives it the name *Pingouin brachiptère*.

THE NAME PENGUIN GIVEN TO THE RAZORBILL.

The Razorbill, *Alca torda*, L., appears also to have been known by the name of Penguin, as mentioned in a map of the Western Isles of Scotland, published at Edinburgh in 1823, and referred to by Mr. Robert Gray in his paper read before the Royal Society, Edinburgh.[56]

It is stated that "the south-west coast of Bernera and Mingulay are remarkably bold precipices, rising perpendicularly from the sea in lofty cliffs of gneiss, which are frequented in summer by innumerable flocks of Puffins, *Razorbill Penguins*, and Kittywakes. These birds disappear early in autumn with their young."

A Welsh name for the Razorbill is *Gwalch y Penwaig*, which is mentioned by Fleming.[57] "*Le Pingouin*" is the appellation given by Buffon,[58] and another synonym is *Pingouin macroptère*, which is applied to the bird by Temminck.[59]

THE SPELLINGS OF THE NAME PENGUIN AND THE VARIOUS CONNECTIONS IN WHICH THEY ARE USED.

The spelling of the name *Penguin* varies considerably, and for the information of our readers we shall give a few of those different spellings, with the connections in which they occur. "*Island named Penguin*."—"Report of the State and Commodities of Newfoundland by M. Anthonie Parkhurst, Gentleman, 1578." [60] "*Island of Penguin*."—"Voyage of the *Grace* of Bristol of M. Rice Iones, made by Siluester Wyet. Shipmaster of Bristoll, 1594." [61] "*Iland called Penguin*," "because of the multitude of birdes of the same name"—In a Latin letter to the

---

[54] "Oiseaux," vol. ix. p. 393.
[55] "Manual," vol. ii. pp. 937–939.
[56] "Proceedings Royal Society, Edinburgh," 1879–80, p. 681.
[57] "British Animals," 1828, p. 130.
[58] "Buffon," Ois, vol. ix. p, 393, pl. XXIX. (1783).
[59] Temminck, "Manual," vol. ii. p. 937–939.
[60] Hakluyt's "Collection of Voyages," London, 1600, vol. iii. p. 133.
[61] *Ibid.*, p. 194.

Worshipfull Master Richard Hakluit at Oxford in Christchurch, Master of Art and Philosophie, his friend and brother." The letter bears date 6th August 1583 in Newfoundland at St. Iohns Port, and is from Steven Parmenivs of Buda.[62]

"*Iland of Pengwin.*"—"A true Report of the late discoueries and possession taken in the right of the Crowne of England of the Newfound Lands by that Valiant and Worthy Gentleman, Sir Humphrey Gilbert, Knight. Written by Sir George Peckham, Knight.[63] The name of the birds is also given as "*Pengwins*" in the same Report.[64] In a notice of the "Voyage of the Ship called the *Marigold* of M. Hill of Redrife vnto Cape Briton (Breton), and beyond to the latitude of 44 degrees and a half, 1593, written by Richard Fisher, Master Hille's Man of Redrife," it is mentioned the "Englishmen land upon Cape Briton. . . . Here diuerse of our men went on land upon the very Cape. . . . And as they viewed the countrey they sawe diuers beastes and foules, as blacke Foxes, Deere, Otters, great Foules with redde legges, *Pengwyns*, and certaine others." [65]

Until quite recently the name *Pin-wing* appears to have been given to the Great Auk in Newfoundland, but this is probably only a corruption of the word "Penguin." [66] Preyer [67] gives the name *Pengwyn* as the Dutch name for the Great Auk, and *Grand Pingouin du Nord* as the French. He also mentions the following synonyms and names—*Plautus pinguis*,[68] Klein, *Pinguinus impennis*,[69] Bonaparte.

In the foregoing we have already given references for the following names, which we merely recapitulate—" le grand Pingoin," " Pinguin," " Penguins," " Northern Penguin," " Le Grand Pingouin," " Pingouin brachiptère," " Penguyn," " Grand Pingouin du Nord," " Plautus pinguis," " Pinguinus impennis."

Apponath.—This is a name which, according to Hakluyt, Iaques Carthier gave to a bird which he found at the " Island of Birds," Newfoundland, and which, from his description, must be assumed to be the Great Auk; but we give his own words, so that our readers may form their own opinions on the subject. The heading of the narrative is as follows: " The first relation of Iaques Carthier of St. Malo, of the new land called *New France*, newly discovered in the yere of our Lord 1534."

---

[62] Hakluyt's "Collection of Voyages," London, 1600, vol. iii. p. 162.   [63] *Ibid.*, p. 172.   [64] *Ibid.*, p. 175.
[65] *Ibid.*, p. 192.   [66] R. Gray, "Proceedings Royal Society, Edinburgh," 1879–80, p. 678.
[67] "Ueber Plautus impennis," Heidelberg, 1862, pp. 14, 16, 17.
[68] "Stemmata avium" Lips., 1759, pp. 36–38, and his : "Prodrom. histor. avium," Lubecæ, 1750, p. 75. And also Historia avium.
[69] Bonnat, "Tabl. Encyl." i. p. 28 (1790).

The beginning of this relation mentions " How M. Iaques Carthier departed from the Port of S. Malo with two ships and came to Newfoundland, and how he entered into the port of Buona Vista." The next part tells : " How we came to the Island of Birds, and of the great quantity of birds that there be," as follows :

" Vpon the 21 of May, the winde being in the West, we hoised saile, and sailed toward North and by East from the Cape of Buona Vista, until we came to the Island of Birds,[70] which was enuironed about with a banke of ice, but broken and crackt; notwithstanding the sayd banke, our two boats went thither to take in some birds, whereof there is such plenty, that unlesse a man did see them, he would thinke it an incredible thing : for albeit the Island (which containeth about a league in circuit) be so full of them, that they seeme to have bene brought thither, and sowed for the nonce, yet are there an hundred-folde as many houering about it as within ; some of the which are as big as iayes (*i.e.* [71] jays) blacke and white, with beaks like unto crowes.[72] They lie alwayes upon the sea ; they cannot flie very high, because their wings are so little, and no bigger than halfe ones hand, yet do they flie as swiftly as any birds of the aire levell to the water ; they are also exceeding fat. *We named them Aporath.*[73] In less than halfe an houre we filled two boats full of them, as if they had bene with stones ; so that besides them which we did eat fresh, every ship did powder and salt fiue or sixe barrels full of them."

" Of two sorts of birds, the one called *Godetz*, the other *Margaulx ;* and how we came to Carpunt.

" Besides these, there is another kinde of birds which houer in the aire, and ouer the sea, lesser then the others, and these doe gather themselues together in the Island, and put themselves under the wings of other birds that are greater ; *these we named Godetz.* There are also of another sort, but bigger, and white, which bite euen as dogs ; *those we named Margaulx.* And albeit the sayd Island be 14 leagues from the maine land, notwithstanding beares come swimming thither to eat of the sayd birds." [74] [75]

The Frenchmen continued their voyage until the 25th June, when we find a notice " Of certain Ilands called the Ilands of *Margaulx,* and of the kinds of beasts and birds that there are found. Of the Iland of *Brion,* and Cape *Dolphin.*"

---

[70] Probably Funk Island.

[71] The original translator into English from the original French has made two mistakes here, as the word he has rendered *iayes* should be *pies* or *magpies* ; and instead of *crowes* ([72]) he should have written *ravens.*

[73] The passage translated in the English version, from which we quote, " *They are also exceeding fat ; we named them Aporath,*" runs as follows in the original French, as it is given by M. Victor Fatio, " Bulletin de la Société Ornitholigique Suisse " (Tome ii. 1er partie, p. 23), " Il y sont excessivement gras, et estoient appelez par ceux du païs Apponath," *i.e.* " They are exceedingly fat, and they were called Apponath (so, not Aporath) by the people of the country."

[74] Hakluyt's " Collection of Voyages," London, 1600, vol. iii. pp. 201 and 202.

[75] The island that Carthier refers to is probably that now known as Funk Island. Polar bears would most likely reach the island upon the ice and not by swimming from the mainland, though they might occasionally swim from passing icebergs.

"The next day being the 25 of the moneth, the weather was also stormie, darke, and windy, but yet we sailed a part of the day toward West Northwest, and in the evening wee put our selues athwart untill the second quarter; when as we departed, then did we by our compasse know that we were Northwest by West about seven leagues and an halfe from the Cape of S. Iohn, and as wee were about to hoise saile, the winde turned into the Northwest, wherefore wee went Southeast, about 15 leagues, and came to three Ilands, two of which are as steepe and upright as any wall, so that it was not possible to climbe them : and betweene them there is a little rocke.

"These Ilands were as full of birds as any field or medow is of grasse, which there do make their nestes, and in the greatest of them there was a great and infinite number of those *that we call Margaulx*, that are white, and bigger then any geese, which were seuered in one part. In the other were onely *Godetz*, but toward the shoare there were of those *Godetz*, and *great Apponatz*, like to those of that Iland that we aboue haue mentioned: we went downe to the lowest part of the least Iland, where we killed aboue a thousand of those *Godetz* and *Apponatz*. We put into our boates so many of them as we pleased, for in lesse then one houre we might have filled thirtie such boats of them. We named them The Ilands of Margaulx." [76]

The reader will notice that, according to Hakluyt, Carthier by his own statement gave the name "*Apponath*" to this bird (this is a mistake, as Hakluyt gives a mistranslation, see note p. 135), and that in the second extract from his narrative which we quote he gives the name as "*Apponatz*." A learned philologist informs us as follows : — "Apponath is clearly an Indian (or Esquimaux) and not a French word. From the very look of the word one could easily tell that it was not French." [77] In another work where an account of Carthier's voyage is given, the name is spelt "*Apponath ;* [78] and [79] Professor J. Steenstrup, who was the first of recent authors to refer to the name, thinks we have good reason to believe it was applied to the Great Auk, and supposes it may be a corruption of the English word *harpooner*, for the French mode of spelling the word varies, it also appearing as "Aponars."

Aponar is the name given to the Great Auk by André Thevet,[80] who, speaking of his voyage along the East Coast of America in 1555, says that at eight degrees from the Island of Ascension there was found a considerable quantity of well-known birds, of which many were large birds with small wings and unable to fly.

[76] Hakluyt's "Collection of Voyages," London, 1600, vol. iii. p. 205.

[77] Professor Steenstrup, writing us 22d March 1885, says, "Probably at least, as the name given by the Eskimos to *Alca torda*, L., is *Akparnok*."

[78] Termaux, "Compans Archives," vol. i. pp. 125, 126.

[79] Apponath and Apponatz are the singular and the plural of the same name. In Old French the plural is often formed by *z* instead of by *s*.

[80] "The singularities of Antarctic France, otherwise called America, and of several lands and islands discovered in our own time," Antwerp, 1558.

He had heard them called *Aponars*, and he adds on this occasion what he had heard told concerning the *Aponars* of Newfoundland. "Moreover, in this Island (Ascension)," he says, "there is found a species of large (birds) which I have heard called *Aponars*. They have small wings, wherefore they cannot fly. They are big and stout like our herons with a white belly, a back black as coal, and a beak like that of a cormorant or other raven. When they are being killed they squeak like pigs. I have wished to describe the bird among the others, because it is found in quantities in an island off Newfoundland right opposite Cape Buona Vista. It has been called the Island of *Aponars*. There is there so great an abundance of them that three large French vessels going to Canada sometimes loaded each their boats at will with these birds on the shore of this island, and all that had to be done was to go ashore and drive them on before into the boats like sheep to the shambles. That is what has caused me to speak of them at such length."

M. Victor Fatio[81] in his French translation of Professor Jap. Steenstrup's celebrated paper (see Appendix, p. 1), says:—

"Next year (1556), Thevet on his way back from South America was driven by contrary winds over against Newfoundland, but he does not seem to have then observed the birds in question. Thevet was a tolerably good cosmographer, and had been sufficiently acquainted verbally by Jacques Carthier with the results of his two first voyages. It is therefore possible, though far from probable, that the ships to which he alludes were those of Carthier's third expedition,—an expedition which seems, however, to have been unknown to him. The Englishmen, Hore and Parkhurst, confirm the fact that these poor birds were surrounded and driven into the boats. Thevet gives us the names of *Godet* and *Margaux*, familiar to the whalers of that time. *Margaux* appears to have signified usually the Gannet (*Sula*). The name *Godet* appears to have been applied in a general way to the black-feathered birds of the genus *Alca*. As to the word *Apponath* or *Apponar* it was used to designate the birds which could not soar in flight, and it was afterwards supplanted by the word *Pinguin*.

"The *Apponar* of the island of Ascension cannot have been the *Alca impennis*, but rather a *Spheniscus*. This would correspond pretty well with what we know of the geographical extension of that species in the south of the Atlantic."

It is perhaps better to leave it an open question, with our present knowledge, whether the name was applied to the Great Auk or not, and it may be as well to state that such an authority as Professor A. Newton does not seem to be quite certain on the point, and thinks it may apply to some other species of Alcidæ.[82]

[81] "Bulletin de la Société Ornithologique Suisse." Tome xi. 1ᵉʳ partie, p. 26.
[82] "Natural History Review," 1865, p. 482.

Mergus Americanus.—Carolus Clusius[83] gives the name of *Mergus Americanus* to a foreign bird of which he had seen no description. He got a coloured drawing sent him by a person described as James Plateau of America, and figured this bird in his work referred to, but the figure is not a good one. However, what is awanting in the picture is made up for by the careful description which he gives, and which the late Dr. John Alexander Smith[84] thought could leave no doubt as to the species. Clusius says, "Rostrum aquilinum et satis crassum, non planum, in quo nulla dentium vestigia expressa apparent; ejus autem partem pronam obliquas quasdam strias habuisse pictura fidem faciebat, et anteriorem capitis paullò supra rostrum partem albâ maculâ insignitam si pictor quidem legitime illam expresserat," &c.

Anser Magellanicus, seu Pinguini.—*Anser Magellanicus, seu Pinguini* was the name used by Olaus Wormius[85] in 1655 for the Great Auk; but he must have had a confused idea about the bird, for though he figures the Great Auk from one got from the Faröe Islands, and which he kept alive for several months (of this excellent figure we have been able to give a reproduction at page 68), yet he describes it as the *Anser Magellanicus, seu Pinguini* of Clusius, while that author describes the true Great Auk as the *Mergus Americanus,* and the Penguin of the Southern Ocean as the *Anser Magellanicus.*

Alca impennis.[86]—Another name, and the one which is likely ever to remain the designation by which the Great Auk will be best known to science, is that given to it by Linnæus—viz., *Alca impennis,* L. Though this is one of its latest names, having only come into use towards the end of last century, still it has been serviceable in clearly separating the Great Auk from the *Spheniscomorphae* of the Southern Hemisphere; and this is no small service, as it has removed the cause of the confusion that is to be found in the writings of ornithologists who lived previous to the general adoption of this name by scientists.

Chenalopex impennis,[87] L. Sp.—This is the name adopted in some of our public museums, and is the name for the Great Auk given in Gray's "Hand-list of Birds

[83] "Exoticorum Decem Libri," Leyden, 1605, lib. v. p. 103.

[84] "Proceedings, Society of Antiquaries of Scotland," 1878-9, p. 96.

[85] "Museum Wormianum seu Historiæ Rerum Rariorum" (Copenhagen), Leyden, 1655, p. 301.

[86] "Linn. Syst. Nat.," i. p. 210 (1766).

[87] "Ueber Plautus impennis," William Preyèr, p. 16; also "Handl. of B.," iii. p. 95, No. 10,773 (1871).

in the British Museum." He mentions that the author of the generic name was Mœhr. (Möhring), who uses it in 1752 ; it was adopted by Vieillot in 1818.

Other names.—Other names for the Great Auk are given by various authors, and the following are those that we have noted in the course of our studies ; but this unfortunate bird was known by so many names that most likely there are others that we have not met with :—*Northern Auk,*[88] English ; *Tossefugl,*[89] Danish ; *Brillenalk*[90] and *fügloser Alk,*[91] German ; *Garefogel*—we give this as a Faröese name at page 124, but it is as the Swedish name we give it here ;[92] Professor Steenstrup writes us, " It is used by some Swedish naturalists ;" *Plautus impennis,*[93] Brünnich ; *Matœoptera impennis,*[94] Gloger ; *Alca borealis.*[95] Unfortunately, Professor W. Preyer is not accurate or reliable in his statements regarding the Great Auk, so we cannot place implicit trust in the names he gives. It is, however, only right to say that this is probably less owing to his fault than to his misfortune in having had to obtain his information second and third hand, the originals not being accessible.

---

[88] "Ueber Plautus impennis," William Preyer, Heidelberg, 1862, p. 17.

[89] *Ibid.* p. 17. Professor Steenstrup, 22d March 1885, informs us " *Tossefugl* is given by some Danish authors as the name for the Solan Goose, *Sula alba*, but never for the *Geirfugl.*"

[90] *Ibid.* p. 17.

[91] "Birds of Europe," H. E. Dresser, vol. viii. p. 563.

[92] *Ibid.* p. 563.

[93] "Ueber Plautus impennis," William Preyer, Heidelberg, 1862, p. 16 ; also "Natur. Foren. Vidensk. Meddel.," 1855, p. 114.

[94] "Ueber Plautus impennis," William Preyer, Heidelberg, 1862, p. 16.

[95] Forster, "Synopt. Cat. Brit. B.," p. 29 (1817).

CHAPTER XIV.

CONCLUSION.—THE PERIOD DURING WHICH THE GREAT AUK LIVED.

WE have been asked repeatedly whether remains of the Great Auk have been found in a fossil state in any geological strata, but so far as we are aware no such remains have ever been discovered, and it has somewhat surprised us to see the change of countenance that has come over our interrogators when we have told them so. To have been found in a fossil state evidently would enhance the value of such remains in the eyes of some persons; but it seems to us that it is more probable that the remains of the Great Auk will become more and more valuable because they have not been found in such a state; and we think this a good illustration of how many extinct birds and animals may have lived abundantly until comparatively recent times and yet have died off, leaving not a single trace that they ever existed. To be preserved as geological specimens certain conditions are requisite, and as these seldom occur, and then are generally confined to limited areas, it is no wonder that we have many missing links in the sequence of animated nature, and that it is difficult for the apostles of evolution to produce all the evidence their opponents require to satisfy them that there are good grounds for the theory they advance.

In the foregoing pages we think we have shown that, with the exception of a very few of the larger and thicker bones, the skeletons of the Great Auks found in the European regions have entirely disappeared, and most probably this has occurred within 2000 years in Britain, and, according to Professor Steenstrup, within 4000 years in Denmark; while in Iceland only a few of the larger bones have been found in deposits that most likely were formed within the last few centuries, and cannot in any case be more than 1000 years old, as it is only that time since the island was colonised. In North America the same experience has been repeated, as with the exception of Funk Island,—the last stronghold of the bird in that region, where several mummy Great Auks were found frozen in the soil,—the remains found in the kitchen-middens have been similar to those discovered in Europe.

These periods are only as moments in days of geological time, and if after the lapse of such comparatively short periods so few bones of a bird are left, we may be sure there is little hope that any remains of the Great Auk will ever be found in geological strata, unless by some unusual and remarkable combination of circumstances they have been brought within reach of the necessary conditions for their preservation in a fossil state.

APPENDIX.

APPENDIX.

I.

RECAPITULATION OF THE VARIOUS INVESTIGATIONS CONCERNING THE DISTRIBUTION OF THE GAREFOWL.

BY PROFESSOR J. STEENSTRUP.[1]

(Translated from the Danish.)

WE must now endeavour to gather up in a few short sentences what we have learned from the preceding accounts of the *Geirfugl's* appearances at the various points mentioned. The better to understand the following summary, the reader will please compare the accompanying map. (See original Paper.)

1. The Geirfugl has never been a really Arctic bird,—that is to say, a bird which chiefly has its resorts and its resting-places within the Arctic Circle. There is, perhaps, not even a single particle of evidence to prove that it has ever been seen within the Arctic Circle, even in casual individual specimens.

The *Geirfugls* that have been observed farthest north are those seven which were killed by an Icelandic peasant on the rock under Lautrum-Fugleberg, as mentioned in his account given to Faber. Yet even these chance stray specimens did not come at all near the Arctic Circle. The most northerly known nesting-places are the Geirfugl-rocks, off the coast of Iceland, between 63° and 64° N. Even if one assumes a nesting-place at Frederikshaab, in Greenland (and that rests only on the doubtful basis of the one problematical young bird mentioned by Fabricius), even it does not lie so far north.[2] Thus

---

[1] As no part of Professor J. Steenstrup's valuable paper on the Great Auk has been translated, as far as we are aware, into English, except what we have given at page 31, we have thought his summing up of the results of his studies well worthy of being brought within reach of British Ornithologists. As the paper referred to was written in 1855, allowance must be made for some slight inaccuracies which have been discovered since we have had fuller information on the subject. But as a whole, the veteran Professor's statements are wonderfully correct.—S. Grieve.

[2] There can be no doubt that it was the young of quite another bird than *Alca impennis* which was seen by Fabricius.—J. Steenstrup, 15th March 1885.

T

the most northerly actually known nesting-place of the Geirfugl does not reach the southern boundary of that zone of latitude which has hitherto been assumed to be the proper home of this bird.

2. Nor has the Geirfugl in later times become an Arctic bird. There are at least no observations in proof of the supposition that it has been chased up from southern parts into these Arctic regions.

3. The home of the Geirfugl, as marked out by its historically known breeding-places, fell along the edge of the upper part (*northern part*) of the North Atlantic Ocean. In this Anglo-Saxon portion of the Atlantic Ocean the breeding-places of the *Geirfugl* formed, as it were, a half-circle at a considerable distance from the coast-line of the mainland or the larger islands. If we begin with the western Geirfugl-rock, off Iceland, as the most northerly point, the semicircle went towards the west down over Funk Island (if over Labrador, is very doubtful), Penguin Island, off the south coast of Newfoundland, the Bird Islands, in the Bay of St. Lawrence, Cape Breton, and not improbably right down to Cape Cod. Towards the other,—*i.e.* the eastern side,—the semicircle passed over the southern Geirfugl-rock, near Westmannöe, and the eastern near Ingolfshofde, the Färöe Islands and St. Kilda.

4. Over all this expanse the *Geirfugl* has been known to breed only on rocks or outlying islands situated at a distance of from two to fifteen miles from the larger islands, or the continuous coastline of the mainland. Now, this can have been the case with the bird from the very first; but we can also quite well suppose this state of matters to have arisen out of recent persecutions of it, so that it has remained only in such distant, not easily accessible, places. If we follow the course of the bird's disappearance, we cannot but be led to the assumption that these outlying islands are only the remains of a former more general distribution of the bird over those islands also that lie nearer the coast, and, it may be, over the coast itself. Certainly, a natural chain of reasoning leads us to assume that its historical decline within the last two hundred years is only a continuation of what was going on for centuries previously, though not then perhaps in the same proportion. A bird so defenceless as the *Geirfugl* cannot be imagined as breeding on the coast of the mainland, or on the most closely adjacent islands, without being in a high degree a spoil for beasts and birds of prey, and above all for man.

5. All the above-mentioned outlying islands are so placed in the ocean and its currents that only very few of them—for example, Funk Island—can be within general reach of the ice-drifts. We have, therefore, no cause to assume a propensity on the part of this bird to live in the neighbourhood of fields of ice.

6. On none of the places on which the *Geirfugl* has been observed within historic times has it been seen in so great abundance as on the islands off Newfoundland. On the whole, the western side of the Atlantic Ocean must be regarded as the chief resort of the *Geirfugl* during historic times, whilst even the earliest notices we have of the *Geirfugl's* appearance on the islands on the eastern side of the Atlantic describe it as being rare and represented by but few individuals.

7. On all the above-mentioned points, it has in the meantime either entirely disappeared, or else come so near extirpation that, so far as our present knowledge goes, there does not seem any likelihood of its existing in any colony of size. On the west *Geirfugl*-rock off

Iceland it still, according to all probability, lives and breeds, although even that colony must be a tolerably small one.

8. The *Geirfugl's* disappearance (and this disappearance must not be regarded as a migration, much less a natural dying out, but as an extirpation) has its chief cause in the devastations wrought by man. With a view to dietary purposes, men have at certain times caught the bird to an extent quite out of proportion to the conditions under which the continued existence of the species was alone possible. Yet the bird, whilst disappearing, has in so far helped to the attainment of a higher object, as it has evidently been for a long space of time one of the means that have essentially facilitated the prosecution of fishing on the Banks of Newfoundland.

Inasmuch as certain of the *Geirfugl's* nesting-places have often been liable to violent natural disturbances, we have in this fact also at least a subsidiary cause of the *Geirfugl's* decrease in and disappearance from a few places—for example, from the *Geirfugl*-rocks, off Iceland.

9. How far down towards the south the *Geirfugl* in former, but now far distant ages, has followed the coasts of North America or Europe, can of course only be decided by similar investigations, and by such finds as that which gave occasion to the present essay. Should bones of this bird come to light in much more southerly latitudes (which I do not look upon as improbable), one may at least regard it as certain that there is now enough known of its history to decide that its disappearance is not attributable to any change in nature, but only to the hand of man.

II.

EPITOMISED TRANSLATION FROM THE GERMAN OF THAT PART OF PROFESSOR WH. BLASIUS' RECENT PAMPHLET ("ZUR GESCHICHTE DER UEBERRESTE VON ALCA IMPENNIS, LINN.;" NAUMBURG, A/S 1884) WHICH TREATS OF THE SKINS AND EGGS.

THE original paper of Professor William Blasius is of such length that we regret being unable to afford space for a full translation of it. The following epitome gives full information regarding skins and eggs of *Alca impennis* which are in foreign collections, information regarding these being difficult for English readers to obtain. Where Professor Blasius refers to skins and eggs in British collections, we generally give the English sources of information, but do not enter into details. We have to acknowledge our obligations to the learned Professor for his permission to give this translation. Where his remarks are condensed we have placed brackets.

LIST OF STUFFED SKINS OF THE GREAT AUK.

1. *Aalholm, Laaland, Denmark.*—Writing to me on the 18th November 1883, Professor Japetus Steenstrup mentions that during the summer of that year he had the opportunity of seeing this specimen. It is the property of Count Raben, and is in his country house at Aalholm, Nysted, Laaland. It is badly stuffed, but as a skin is well preserved. No previous mention appears to have been made of this specimen in the literature of the Great Auk.

It was a Danish count, F. C. Raben, who went along with Faber and a Danish botanist, Mörck, in that dangerous voyage (29th June to 2d July 1821), when the subsequently submerged Geirfuglasker was visited for the last time, and when also a visit was paid to the skerry off Reykjanes, called the Grenadier's Cap, which also has not been visited again up to this time. In both cases it was Count Raben alone who effected a landing, and that, too, at the peril of his life. Writing to me on 25th November 1883, Professor Newton says that if the present Count Raben is the representative or actual descendant of Faber's fellow-voyager, and the Aalholm specimen belonged originally to him, then he must probably have obtained it after his return from Iceland, for Faber says distinctly that during his three years' stay in Iceland (1819–21) he could not procure any specimen of the Great Auk.

[But Blasius says that Count Raben may have got his specimen after he had parted company with Faber. He thinks that the Aalholm specimen may be one of the two which Wolley and Newton (*Ibis*, 1861, p. 387) mention on the authority of Sigurd Jonsson as having been killed with a sail-pole on a low-lying rock off Hellirknipa, between Skagi and Keblavik, a few days after the end of Faber's sea-trip. Faber, Blasius points out, could have had no opportunity of learning about these two specimens. He says that as these two specimens, whose subsequent history has never been clearly elucidated, were, on the authority of Sigurd Jonsson, skinned in the same way as Arctic foxes and other mammals, namely, by having a hole pierced transversely through the legs, if it could be found out that the Aalholm specimen has been skinned in the same way, it would serve to confirm his conjecture.]

I shall not, however, conceal the fact that Newton brings forward a very probable explanation of the subsequent history of these two specimens. Jonsson states the skins came into the hands of Asgrimur Saemonsson, an inhabitant of Eyrarbakki. Now, it was exactly from Eyrarbakki that, in 1823, the skins of two old birds in breeding plumage were, according to Faber (*Isis*, 1827, p. 683) sent to the Royal Museum of Copenhagen, the information given regarding them being that they were killed by a boy with a stick on a skerry near that place. Newton, therefore, would hold the two pairs of skins to be identical. But even if this conjecture be right, there was, according to Newton's investigations, a *third* skin that figured at that time, for he says (*Ibis*, 1861, p. 387), "Some persons to whom we spoke said that Faber had got three specimens." By this it can only be meant, seeing that Faber actually got no skins at all in Iceland, that they sought to sell three skins to him. This third bird may have been a third one killed near Hellirknipa, or perhaps the specimen caught in the year 1818 on South Island, mentioned not only by Kjärbölling (Danmark's "Fugle," 1856, p. 415), but also before him by Faber (*Isis*, 1827, p. 682). At any rate, it is possible that Count Raben got this third specimen afterwards, as mentioned by some eye-witnesses, during his stay in Iceland at that time.

At all events, it is in the highest degree probable that the specimen originally came from Iceland.

2. *Aarau, Switzerland.*—[This is a fully-coloured specimen, and was presented to the Museum of Aarau by Counsellor of State Frey-Hérosée, sometime about the year 1865. Frey-Hérosée bought it in 1842 or 1843 from the father of Dr. C. Michahelles, who had died at Nauplia in 1835, on his paying the price of 80 florins which was asked for it by a friend of Dr. C. Michahelles. Victor Fatio thinks that this is one of the specimens described by Michahelles in *Isis* (year 1833, page 650). But Blasius thinks that the friend of Dr. C. Michahelles, remaining in ignorance of the latter's death in 1835, sent it to his address about 1840. In that case it would be one of the three skins which, as Wolley and Newton ascertained (*Ibis*, 1860, p. 390; "Journal für Ornithologie," 1860, p. 327), were sold in August 1840 or 1841 along with the body of a bird and a number of eggs by Factor C. Thaae to S. Jacobsen, the other two being at present probably in Bremen and Oldenburg. But certainly, Blasius says, the Aarau specimen is of Icelandic origin, as all Michahelles' specimens came thence.]

3. *Abbeville, France.*—[The Musée de la Ville possesses a specimen coming from the collection of De la Motte. (See A. Newton in "Nat. Hist. Review," Oct. 1865; Journal für Orn., 1866, p. 104, and *Ibis*, 1870, p. 258.) Doubtful whence De la Motte got it, C. D. Degland ("Ornithologie Européene," first edit., vol. ii. p. 529, date 1849; and "Naumannia," p. 423, date 1855), says that from forty to fifty years before his time three

specimens were killed at Cherbourg, one of which found its way into De la Motte's collection. But A. Newton (in "Nat. Hist. Review," Oct. 1865, and "Journal for Orn.," 1866, p. 404) thinks that he has clearly proved that the Abbeville specimen came from Copenhagen, and in all probability originally from Iceland. .Writing to Blasius, 17th Nov. 1883, he says : "The specimen in the Abbeville Museum was certainly not killed at Cherbourg ; it was got in Iceland, and sent in the year 1831 to Monsieur De la Motte from the Royal Museum in Copenhagen, as I have learned from my good old friend Reinhardt. I thoroughly distrust Degland's story about the birds killed at Cherbourg." Dresser ("Birds of Europe," vol. viii. p. 565) also discredits Degland's account.]

4. *Amiens, France.*—[The Musée de la Ville here possesses a specimen which A. Newton saw himself (*Ibis*, 1870, p. 258 ; "Bull. Soc. Ornith. Suisse," tome ii., partie 2, p. 151), and that, as he writes to Blasius, as late as 1862, so that this specimen assigned to Amiens can by no means be the same as that which Capt. A. Vouga of Cortaillod got about the year 1838 by way of Amiens. Prof. A. Newton, writing to Blasius, thinks, but is not quite sure, that this Amiens specimen came from the collection of A. Delahaye, formerly Director of the Natural History Museum of St. Omer, who got it originally from Mechlenburg at Flensburg. Thus this specimen is probably of Icelandic origin.] See also p. 79.

5. *Amsterdam, Holland.*—This specimen, which is to be found in the Museum of the Royal Zoological Society, was first mentioned by Champley ("Ann. & Mag. Nat. Hist.," 1864, p. 235). It is not, as I first conjectured, the specimen which the Mainz Museum disposed of by way of exchange to G. A. Frank, dealer in zoological wares, Amsterdam, but was, as Herr G. A. Frank, London, the son of G. A. Frank, has informed me, obtained direct from his father on a different occasion ; namely, in 1843. Professor A. Newton saw this skin as early as 1860.

6. *Berlin, Germany.*—[Is in the Royal Zoological Museum. Given in Lichtenstein's "Nomenclator Avium," 1854, p. 105, with the reference "Polar Sea." Preyer ("Journal für Ornithologie," 1862, p. 78) says that it bears the inscription : "Alca impennis, Linn., Polarmeer. Reinhardt." Prof. J. Cabanis of Berlin, writing to Blasius, says that it was in the Museum before his time, *i.e.* before 1838, and that it probably found its way into the Museum through the close intimacy subsisting between Lichtenstein and Reinhardt (senior), Director of the Royal Museum of Copenhagen, whose name appears on the inscription. Like all the specimens coming from Copenhagen, it must be of Icelandic origin.[1] According to Prof. W. Preyer ("Plautus Impennis," Diss. 1862, p. 11, and "Journal für Ornithologie," 1862, p. 119) it has on the point of the beak eight cross-furrows above and twelve below.]

---

[1] From some manuscript notes in his possession regarding the history of *Alca impennis*, Professor Steenstrup has kindly furnished us with the following information : "Professor Lichtenstein, in a letter dated 15th March 1831, acknowledged receipt of the said skin." Professor Steenstrup adds, "Our museum had got it from Iceland the same year."—S. Grieve.

7. *Boyne Court, Essex, England.*—[Prof. Newton is informed by a letter from Mr. A. D. Bartlett that the specimen which belonged to Mr. Hoy, who died more than 40 years ago, is now in the possession of Mrs. Lescher, a sister of Mr. Hoy's.]

8 and 9.—*Brunswick, Germany.*—[One is the property of the Museum ; the other is a loan from a private party. The former was bought from Frank certainly before May 1842, and probably between 1830 and 1840. The latter can be traced back to the possession of Baron von Pechlin, Danish deputy to the old Diet of the German Confederation. Cross-furrows of the former $\frac{7}{10}$, of the latter $\frac{7}{12}.\frac{7}{13}$. Both probably of Icelandic origin, as can be told especially from the manner in which they have been prepared.]

10. *Bremen, Germany.*—A specimen in a good state of preservation in the Municipal Collections for Natural History and Ethnography. It was bought at the time of the Bremen Congress of German Naturalists in the year 1844 by Dr. Hartlaub from Salmin of Hamburg, a dealer in zoological wares, for the price of £6. In all probability this is one of the three skins mentioned by Wolley and Newton as having been got on Eldey island near Iceland in the year 1840 or 1841,[1] the other two being now in Aarau and Oldenburg (or perhaps Kiel). But perhaps it is the specimen that Frey Hérosée sent away about this time in exchange to Hamburg.

11 and 12. *Breslau, Germany.*—Two specimens in the Zoological Museum of the University, to which attention was first called by Alexander von Homeyer in the "Journal für Ornithologie," 1865, p. 151. He says they are the skins of aged birds, and are presumably to be regarded as a pair (♂ and ♀). The furrows on the points of their beaks are, he says, respectively $\frac{10}{11}$ and $\frac{8}{11}$. Regarding their origin, Prof. Dr. A. Schneider, the present director of the said Museum, wrote to me on the 24th Nov. 1883 : " No documents remain relating to the origin of the two *Alca impennis* in our possession. According to the account of the conservator, which rests on oral tradition, they were both probably bought from an itinerant dealer between 1830 and 1840. I conjecture that the dealer in question would be Platow, who was well known in German towns as a museum proprietor and a dealer in zoological wares. Pässler (in the "Journal für Ornith.," 1860, p. 59) says that Platow is known to have sold two eggs of this rare bird. The family Platow belongs to Freiburg in Silesia, and it is very likely that the itinerant dealer who sought all round wares to buy, in order again to sell them all round, would bring back from one of his tours two Great Auks for the collections of Breslau, the capital of his native province. Heinrich Platow, the son of old Platow, continued the business in a similar manner after his father's death, and was, for example, on his journeys with his museum in the spring of 1879 in Brunswick, in the summer in Witten, Bochum, &c., in the winter following in Solingen, &c.

---

[1] Professor Steenstrup informs us " that the statement is doubtful, and that he rather thinks the skin was got in 1844." He then goes on to say that " he prefers a certain tradition met with here (Copenhagen), that the skin sold to Bremen in 1844 belonged to one of our last individuals (one of those preserved in spirits) got in 1844."—S. Grieve.

13 and 14. *Brighton, England.*—There were two fine specimens in the collection of George Dawson Rowley, who died on 21st November, 1878 (see Newton in *Ibis*, 1870, p. 259). The collection now belongs probably to his widow (*should be his son*). One of them belonged, up to 1869, to the collection of Count Westerholt-Glikenberg, of Wester-holt, in the district near Münster, and Altum ("Journal für Orn.," 1863, p. 115) mentions it as being his. Herr G. A. Frank of London finds in the business books of his late father, G. A. Frank of Amsterdam, that the latter in 1869 bought it for 10 Louisd'or out of the collection of Count Westerholt-Glikenberg, and soon sold it again to George Dawson Rowley. Rowley's second specimen was bought in 1868 from Gardner of London, who got it in 1848 from Lefèvre of Paris. It is probable that the collection of the deceased Mr. Rowley (which it may be said contained also six eggs) is preserved in his mansion of Chichester House, East Cliff, Brighton.

15. *Brussels, Belgium.*—An old specimen in breeding plumage, first mentioned by A. Newton (*Ibis*, 1870, p. 259), is to be found in the Musée Royal d'Histoire Naturelle. It was bought when the Musée was under the management of Viscount Bernard du Bus de Ghisignies (*ob.* 1874).

16. *Cambridge, England.*—Champley ("Annals and Mag. of Natural History," vol. xiv., p. 235, year 1864) seems to be the first to mention this specimen. It is in the University Museum of Zoology. Some time ago Prof. A. Newton informed me that some of the bones from its extremities were taken to complete the Cambridge skeleton, which, as well as three eggs, belong to the collection of the brothers A. and E. Newton.[1]

17. *Chalon-sur-Saône, France.*—[Dr. B. F. de Montessus, a well-known ornithologist of this town, bought from Dr. L. W. Schaufuss of Dresden the skin of an *Alca impennis,* which Schaufuss had in January 1873 advertised for sale by means of a circular. As Dr. Schaufuss tells Blasius by letter, Montessus had previously bought from him numerous birds for his beautiful collection (mentioned in "Naumannia" 1855, I., p. 112), so that there is good reason for believing that the auk then acquired by Montessus has been permanently in-corporated in his collection. He got into the bargain two skins of the *Alca torda* in order to improve it, for it was only a badly stuffed bird again turned into a skin. Its origin is doubtful. Schaufuss thinks it is the one that once belonged to C. E. Götz of Dresden (see Al. Naumann, "Bull. Soc. Ornith. Suisse," tome ii., partie 2, p. 148). But this, says Blasius, is very unlikely. Götz's specimen, according to the joint testimony of G. A. Frank of Amsterdam and Spencer F. Baird of Washington (cited by A. Newton, *Ibis*, 1870, p. 269) was sold to the Smithsonian Institution in Washington. Perhaps it is the specimen which was left by the apothecary Mechlenburg of Flensburg in his collection at his death

---

[1] Professor Newton, in a letter to the author, dated 17th Sept. 1884, says: "The existence of the stuffed skin at Cambridge was first recorded, some forty or forty-five years since, in the printed Catalogue of the Museum of the Cambridge Philosophical Society, whence it was transferred to that of the University. At page 127 of his treatise Professor Blasius retracts what he states above, and says the skin was first mentioned by Jenyns in 1836." (Catal. Coll. Mus. Camb. Philosp. Soc. p. 15.)—S. Grieve.

in 1861. Up to this time Flensburg has always been credited with a specimen ; but there is now no skin or stuffed specimen whatever of the *Alca impennis* there.[1]]

18. *Clungvnford, Shropshire, England.*—[A Mr. Rocke here, now dead, possessed a specimen. Since his death, which took place not so long ago, it is probably, says Newton, writing to Blasius (26th Nov. 1883), still in the hands of some member of his family. This specimen has often changed its owner. G. A. Frank sold it, along with another specimen of the same kind, in 1835, to the Museum of Mainz (Hellmann in the "Journal für Ornith.," 1860, p. 206), but got it back again in 1860 in exchange for the skin of an Indian tapir (W. Preyer, "Journal für Ornith.," 1862, p. 78), a fact corroborated to Blasius by Herr Frank of London from the books of his deceased father, Herr G. A. Frank of Amsterdam. Newton tells Blasius that immediately after Frank sold it to Gould, and that shortly after, perhaps in that same year, 1860, it passed from Gould's hands into those of the late Mr. Rocke.]

Copenhagen.—See *Kopenhagen* under K.

19. *Cortaillod, Neuchatel, Switzerland.*—In the private collection of Captain A. Vouga of Cortaillod is a beautifully preserved old bird in summer plumage, of which Victor Fatio has given full details ("Bulletin Soc. Ornith. Suisse") on the information of Captain Vouga. He has also given at the end of his paper a coloured plate of the bird after a water-colour done by a son of the Captain's. In 1868 Vouga had already been some thirty years in possession of it. According to him it was brought preserved in salt by some whale-fishers to a port in the north of France. It was stuffed in Amiens, and sent by the dresser of it to a business friend of his, from whom in turn Vouga got it. One is inclined to suppose that the French whale-fishers brought it from the coasts of Newfoundland ; though it may just as well have been caught on some Icelandic skerry, and then salted to serve for provisions.

19A. *Darmstadt, Germany.*—[There is a fac-simile in Darmstadt, of which only the head is genuine.]

20. *Dieppe, France.*—C. D. Degland ("Ornithologie Européenne," first edit., 1849, tome ii. p. 529), following the information given by Hardy in his catalogue of birds observed in the department of Seine-Inférieure ("Annuaire de l'Association de la Basse Normandie," 1841, p. 298), states that in the first decade of this century, in the month of April of two different years, two Great Auks were found on the shore near Dieppe. One of them was killed, the other found dead. One of them found its way into Hardy's collection. Champley ("Ann. and Mag. Nat. Hist.," vol. xiv., 1864, p. 235) mentions both a stuffed bird and an egg as existing in Hardy's collection (which is now presumably in the Musée de

---

[1] Olphe-Galliard (Contributions á la Faune ornithologique de l'Europe occidentale. Fascicule I. 1884, p. 27) mentions this specimen independently of my accounts, and evidently on the ground of direct information, so that its existence in Dr. de Montessus's collection may now be regarded as certain.—W. Blasius.

la Ville), and this seems tolerably certain; but the alleged origin of this specimen of Hardy's is very much doubted by Professor Newton, with whom Dresser ("Birds of Europe," vol. viii. p. 565) agrees. It is to be presumed that the Dieppe specimen, like the Abbeville one, is of Icelandic origin. My earlier communication regarding the origin of this specimen, in the "Third Yearly Report of the Brunswick Association for Natural History," must consequently in all probability be corrected.[1]

21. *Dresden, Germany.*—There is in the Royal Zoological Museum, besides the egg out of Thienemann's collection, a well-preserved specimen of the Great Auk (mentioned also by Preyer in the "Journal für Ornithologie," 1862, p. 78). Writing, 23d Nov. 1883, the director of that Museum, Herr Hofrath Dr. A. B. Meyer, has kindly furnished me with the following particulars : " Our *Alca impennis* belongs to the old contents of the collection, and is, so far as I know, first mentioned by Reichenbach in his manual, 'Das Königliche Naturhistorische Museum,' 1836, p. 22. Of its origin I know nothing. Reichenbach has given a representation of it in his 'Vollständige Naturgeschichte.' It must have been a fully grown bird in completely coloured plumage."

22. *Dublin, Ireland.*—[See Newton, " Natural History Review," October 1865. This specimen is especially interesting from its plumage. Edmund de Selys-Longchamps says (" Comptes-rendus des Séances de la Société Entomologique de Belgique," 1876, 7th Oct., p. lxx.) that of all the many specimens examined by him in the different museums of Europe this is the only one in winter plumage.]

23. *Durham, England.*—This specimen is in the University Museum. It is an old bird. It was first mentioned by Champley in the " Annals and Magazine of Natural History," 1864, vol. xiv. p. 235. This specimen was formerly in the collection of Mr. Gisborne, who got it from one Reid of Doncaster, who had got it from F. Schultz of Dresden (presumably the same as F. Schulz of Leipzig).

23A. *Flensburg, Germany.*—[Prof. A. Newton had stated, on " indubitable authority," that there were at Flensburg, in the possession of Mechlenburg, an apothecary, 8 skins and 3 eggs. In order to obtain certain information about some of these, Blasius put himself in communication with Herr W. Toosbüy, the head burgomaster of Flensburg, whose reply showed that Flensburg in all probability is no longer to be credited with any specimen whatever of the *Alca impennis*. After consulting with Herr Bentzen, the grandson of the deceased Herr Mechlenburg and his successor in the business, he wrote thus on the 30th November 1883 :—"There have certainly been some specimens of the *Alca impennis* in the collection of the late Herr Mechlenburg ; though whether he was possessed of eight is very doubtful. Already in his lifetime he had disposed of the most of these specimens,

[1] In Mr. Hardy's catalogue, 1841, this specimen is not mentioned as being in his collection. Hence it must have been bought subsequently. Professor Newton saw the skin and egg at Hardy's house in 1859. Hardy died 31st October 1863.—W. Blasius.

[2] There are specimens in winter plumage at Copenhagen and Prague.—S. Grieve.

and we find among his memoranda that 'a skin of the *Alca impennis*, but without the feet and breast-plumage, was sold for 1000 marks Schleswig-Holstein currency (= £60) to Siemsen, a merchant in Reykjavik,' Iceland, and that 'Mr. Robert Champley bought an *Alca impennis*, with an egg, for £120.'[1] What became of the other skin that remained after Herr Mechlenburg's death the heirs do not know. No part of Mechlenburg's effects came into the possession of the town of Flensburg; at the time of his death no scientific man appears to have troubled himself about the collection, and after the lapse of such a long time it would be very difficult to acquire further information."

Accordingly, for the time at least, the name of Flensburg must be deleted from our list. One specimen of *Alca impennis* is mentioned in the catalogue of birds and reptiles which were exposed for sale in the year 1861, after the death of Herr Mechlenburg. Blasius supposes that this is the specimen now at Chalon sur Saône, in the possession of Dr. de Montessus, it having passed through several persons' hands in the interval.

Blasius says that it is interesting to note how Herr Carl F. Siemsen of Reykjavik, through whose hands, at the time when the *Alca impennis* still lived in Iceland, no fewer than 21 birds and 9 eggs passed (see Newton, *Ibis*, 1861, p. 392), bought back at a later time a faulty specimen. Perhaps, he says, this specimen really found its way back to Iceland, and if so is now the only specimen there. Further information as to whether it is really in Reykjavik is very much to be desired.[2]]

24. *Floors Castle, Roxburgh, Scotland.*—[Particulars of this bird and conjecture as to its origin received from Mr. Symington Grieve, " who has made careful studies of the history of the Great Auk, and intends soon to make them public."—Also Professor A. Newton is able to confirm, through information acquired by him, the fact that the specimen has been in Floors Castle upwards of forty years; he, however, put forward the conjecture that the skin was not obtained in Iceland, but was bought from a London dealer. The present Duke of Roxburghe believes that it was bought in Edinburgh between 1830 and 1840.[3]]

25. *Florence, Italy.*—[In the " Museo Zoologico del R. Istituto di Studi superiori." First mentioned by Champley ("Ann. and Mag. Nat. Hist.," 1864, vol. xiv. p. 235). Selys-Longchamps says that it is in summer plumage. Newton (*Ibis*, 1869, p. 393) traces it up through Dr. Michahelles and Frank to Mechlenburg of Flensburg. Professor Giglioli informed Blasius that it was bought for the Florence Museum probably between 1830 and 1833, certainly not later than 1833. It bears on its stand only the old inscription, " Schulz Schaufuss."] See also p. 80.

---

[1] Mr. Champley informs me that the sum he paid for the skin and egg was £45, not £120.—S. Grieve.
[2] In a letter to me, dated 15th April 1884, Professor Newton says that apparently this skin now exists in the Central-Park Museum, New York.—W. Blasius. Professor Steenstrup, writing us, 15th March 1885, regarding the whole of the last paragraph under Flensburg, remarks, " I think there is an error here in some way."—S. Grieve

[3] This specimen has been described in a paper read before the Royal Physical Society, Edinburgh, by Mr. John Gibson, 18th April 1883. (" Proceedings Royal Physical Society, Edinburgh," 1883, p. 335.)—S. Grieve.

26. *Frankfurt-on-the-Main, Prussia.*—Alexander von Homeyer was the first to call public attention to what he called a very fine specimen of the *Alca impennis* in the Senckenbergian Museum here. ("Journal für Ornithologie," 1862, p. 461.) Herr Dr. August Müller, of the Zoological Institute, "Linnæa," in Frankfurt, informs me that the custodier of the Museum says that nothing is known as to whence, when, and from what hands it came into the possession of the Museum, and that the sex of the bird and its condition when caught are likewise unknown. "It is," he says, "a completely coloured but not very well preserved specimen, and is ticketed only 'Northern Europe.' The Museum has no eggs or bones."

27. *Gotha, Germany.*—A communication by Dr. Hellmann, the former director of the Ducal Museum here, to the "Journal für Ornithologie" for 1860 (p. 206), calls attention to a very fine specimen under his charge, which was bought in 1835 from Frank, dealer in zoological wares, Leipzig. I myself saw it there in 1882. It wears breeding plumage. Professor Burbach tells me that it was bought for 20 thalers, and that the source whence it was acquired is strangely given as Ramann.[1]

28. *Graz, Austria.*—Professor A. Newton, in the *Ibis* for 1870, p. 257, was the first to call attention to this one, which is to be found in the Natural History Museum of the Joanneum. In answer to a request of mine for information regarding it, Dr. Sigmund Aichhorn, the superintendent of the Museum, thus wrote to me (24th November 1883) :— "Our specimen is stuffed, and is in summer plumage. It is tolerably well preserved, and was sent in 1834 by Professor Reinhardt of Copenhagen to Herr Josef Höpfner, a landed proprietor at Althofen in Carinthia, who in turn made a present of it to the Joanneum. I cannot tell whether it is a male or a female bird."

Before I got this information from Dr. Aichhorn, Professor A. Newton had informed me that he had found in his manuscript notes the remark that in 1833 a specimen was sent from the Royal Zoological Museum in Copenhagen to Höpfner in Klagenfurt. I conjectured that this presumed specimen at Klagenfurt was identical with the Graz specimen, and my conjecture was in a measure confirmed by a letter from Herr J. L. Canaval, director of the Natural History Museum at Klagenfurt. He says (6th Dec. 1883)—"The Museum here has no *Alca impennis*. In 1833 the Museum did not exist, having been founded towards the end of 1847 ; Count Egger, however, had a collection which he gave up to form the nucleus of the Museum. Herr Höpfner at that time gave the Museum several very interesting contributions. I could, however, learn nothing of an *Alca impennis*."

Were there, then, two Höpfners, one in Graz, the other in Klagenfurt, both of whom received Great Auks from Copenhagen ? A letter from Professor Aug. von. Mojsisovics of Graz (9th December 1883) decides this. He says : "There was in our province (Carinthia)

---

[1] It has in the meanwhile become doubtful whether this specimen really came from Frank's hands. There still lives in Gotha the person who sold the bird through the agency of one Herr Ramann, and he has informed Professor Burbach that he got the skin, along with other objects of natural history, from Greenland, they having been procured by a missionary who had been there.—W. Blasius.

only one ornithologist of the name of Höpfner, who gifted his very rich collection of skins to the Joanneum of this town. Among this was the *one* specimen of the Great Auk,—the pride of the zoological collection."

The Icelandic origin of the Graz specimen is thus clear.[1]

29. *Hanover, Germany.*—Cabanis ("Journal für Ornithologie," 1862, p. 78) was the first to mention this one in the Provincial Museum of Hanover. It is in summer plumage. The custodier of the Museum, Herr Braunstein, tells me that it has been cut up through the belly, and not under one of the wings. He also states that it was formerly (some time previous to 1850) in Clausthal, in the Harz. As, according to information given me by Professor Newton, a specimen was a long time ago sold by Frank, the dealer in zoological wares, to Clausthal Museum, with which the specimen in Hanover is clearly identical, the Icelandic origin of this specimen is very probable.

30. *Hawkstone, Shropshire, England.*—There is here one in the collection of Viscount Hill. According to Newton, it was bought in 1838 from Gould. Champley is the first to mention it ("Ann. and Mag. of Nat. Hist.," 1864, vol. xiv. p. 235).

31. *Kiel, Germany.*—In the Zoological Museum of the University. (See "Third Yearly Report of the Brunswick Association for Natural History," p. 94.) It was very probably bought in 1844 by Professor Behn, along with other articles of natural history, for Kiel Museum, by means of a large sum of money, which in that year the "Prelates and Proprietors" of Schleswig-Holstein voted out of a common fund for the purposes of the Museum. I once thought that this might be the Great Auk seen and caught in Kiel harbour during the last decade of last century; but this can hardly be. I would rather suppose it to be one of Mechlenburg of Flensburg's eight, and indeed that one whose present abode has not yet been cleared up. Or it may have been acquired from Salmin, a dealer in zoological wares in Hamburg, who in 1844 sent Great Aks to Bremen and Oldenburg, and who had at that time received some skins direct from Iceland, and also was in possession at one time of one formerly belonging to Dr. Michahelles.[2]

32. *Köthen, Germany.*—In the Ducal, formerly Naumann's, Collection. First mentioned by Preyer ("Journal für Ornithologie," 1862, p. 78). Dr. E. Baldamus writes to me that it came from Copenhagen.[3]

---

[1] In September 1884 I was myself able to examine this specimen. It bears the summer plumage, with two very distinct spots on the breast. The furrows on the bill are $\frac{7-8}{10\ 11}$. The preparation is apparently by means of sewing in the middle of the body. There are traces of moth-eating on the left white eye-spot and on the right side, on the wing and under jaw.—W. Blasius.

[2] Professor Steenstrup writes us 15th March 1885:—"If really purchased in 1844, it might perhaps be the second of these two Garefowls got in 1844, but traditionally I never heard that mentioned."—S. Grieve.

[3] Apart from Naumann's own description, Pässler was afterwards the first to mention this specimen. —W. Blasius.

33 and 34. *Kopenhagen (Copenhagen), Denmark.*—[There are in the Royal University Zoological Museum of Copenhagen, besides the preserved intestines of two Great Auks, an egg and numerous bones, two stuffed specimens,—one of Icelandic origin, in summer plumage; the other, by many supposed to have come from Greenland, in winter plumage. That this latter, however, came from Greenland, is a supposition that hangs on a very slender chain of evidence. The late Professor Reinhardt of Copenhagen, writing to Professor Newton, holds that it has never been proved that this bird in the winter plumage originally came from Greenland.[1]

But at the same time the existence of the Great Auk on the coasts of Greenland is a historical fact. One caught in 1815, near Fiskernaes, was afterwards in the possession of a Herr Heilmann; and one caught at Disco, in 1821, was in 1824 in the possession of Herr Benicken, of Schleswig, who writes in March of that year to the *Isis* (p. 886 f.) regarding it. He describes it as a bird in winter plumage, and it is this fact which forms the chief ground for identifying this bird of Benicken's with the one now at Copenhagen.]

35. *Leeds, Yorkshire, England.*—This one is now in the Museum of the Philosophical Society of Leeds, who got it on loan from Sir Frederick Milner, the son of Sir William Milner, in whose collection it was for many years. Sir William was led by Mr. Graham, a bird-stuffer of York, to believe that it came from the Hebrides (see Newton in *Ibis*, 1861, p. 398), but Newton, writing to Blasius a few days back, says that he has every ground for believing that this specimen was originally given by Gardiner to Mr. Buddicorn, and that it comes from Eldey, in Iceland.

36. *Leighton, Wales.*—[This one belongs to the collection of Mr. Naylor, an intimate friend of Gould's. It was first brought to light by Newton in the *Ibis* (1870, p. 258). Newton, writing to Blasius on 12th December 1883, says that old Leadbeater informed him that Naylor bought it from him in 1861. He (Leadbeater) had got it from Parzudaki of Paris.]

37. *Leipzig, Germany.*—At the request of Professor Newton, according to whom a specimen was sold by Frank long ago to the Museum of Leipzig, I wrote, making inquiries, to Professor R. Leuckart, the Director of the Museum of the University there. On the 23d November 1883 I received from him the following interesting information:—"Our Museum possesses a fine specimen of the *Alca impennis*. It is a question whether it is the same specimen that Frank sent to Leipzig. It is ticketed 'Iceland: D. Uckermann.'

---

[1] This account rested on a misunderstanding, and has been retracted a short while ago by Professor Newton. Not only Reinhardt believed that it came from Greenland, but Newton, in his article on the Birds of Greenland in the " Arctic Manual," has expressly recognised the fact.—W. Blasius.

Professor Steenstrup, in a letter dated 16th March 1885, refers to the specimen in winter plumage preserved at Copenhagen as follows :—"Our individual in winter dress I have no doubt is identical with Benicken's, and with Heilmann's got in 1815, and is the same as the individual seen by Faber in Copenhagen before his voyage to Iceland, because it was the only known *Alca impennis* from Greenland. Reinhardt, senior (the father), bought this not very good specimen at a price three times higher than the price paid for the excellent specimens from Iceland offered to the museum in the same years."—S. Grieve.

Uckermann is the name of a man who has presented numerous interesting objects to our Museum. Besides, I find it already inserted, in the catalogue of date 24th February 1841, among the animals which Pöppig then arranged as the nucleus of our collection. I have not come upon any further information regarding it."

Naumann, in his " Naturgeschichte der Vögel Deutschlands," vol. xii. p. 646 (1844), mentions that twenty-five years previously a skin, bought for a handsome price, had come to Leipzig from Iceland by way of England. Probably that skin is identical with the specimen now in the Museum of Leipzig, though by it may only be meant some skin bought by old Frank of Leipzig, the dealer in zoological wares,—father of the Amsterdam Frank, and grandfather of the London Frank.

38. *Leyden, Holland.*—This specimen in the Zoological Museum here was first made known to the world, so far as I am aware, by Sclater in the " Annals and Magazine of Natural History," 1864, vol. xiv. p. 320, on the information of Herr Hartlaub. I think I remember hearing Herr Frank of London say that his father, Herr Frank of Amsterdam, sold it to the Leyden Museum, to which, indeed, he furnished many other articles of natural history. Schlegel, in the " Muséum des Pays Bas " (Urinatores, p. 13, April 1867), thus describes it:—"Adult: wing, 5″ 11‴; tail, 2″ 11‴; bill length, from the front, 36‴; height, 18‴; breadth, 8½‴; tarsus, 21‴; middle toe, 33‴."

39. *Lille, France.*—This one is in the Musée d'Histoire Naturelle de la Ville. It belonged formerly to C. D. Degland. For further particulars see Olphe-Galliard in *Ibis* (1862, p. 302). Herr G. A. Frank saw it there in 1883.

40. *Lisbon, Portugal.*—[In the Museum Nacional. See *Ibis*, 1868, p. 457, and *Ibis*, 1870, p. 450.]

41. *Longchamps*, near *Waremme, Belgium.*—[See *Ibis*, 1870, p. 259, and p. 450.]

42 and 43. *London, England.*—[These two are in the British Museum. For the history of both see " Natural History Review " for 1865.] See also page 78.

44. *London, England.*—[This one belongs to Lord Lilford, who, as Newton informs me by letter (17th November 1883), got it after the death of his brother-in-law, Mr. Crichton. Public attention was first directed to this specimen by Professor Newton in *Ibis* (1870, p. 258). It is at present in the offices of the British Ornithologists' Union, 6 Tenterden Street; but in all likelihood it will be removed, when it is found convenient, to Lord Lilford's seat, Lilford Hall, Oundle, Northampton.]

45. *Lund, Sweden.*—[In the Zoological Museum of the University. First brought under public notice by Professor Newton, on the authority of Wolley. It is marked " Greenland, 1835: from Reinhardt and Nilsson ; " but Blasius has the authority of Professor Quennerstedt, the Director of the Museum, for saying that it may be regarded as certain that this bird is of Icelandic origin, since it was presented, in 1835, by Herr Reinhardt,

Counsellor of State, Copenhagen, through the agency of Professor Nilsson, who lately died at Lund at the advanced age of ninety-six.]

46. *Milan, Italy.*—[This one is in the very rich private collection of Count Ercole Turati, who died a few years ago. Regarding this specimen, see *Ibis*, 1870, pp. 450, 609.[1]]

47. *Mainz, Germany.*—[In the Town Zoological Museum here there is now only one of the two examples which Herr Frank, of Amsterdam, the dealer in zoological wares, sent to Mainz. The other found its way latterly, after some changes of hands, into the possession of the late Mr. Rocke, of Clungunford.]

48. *Metz, Germany.*—[In the Town Museum. This is the specimen which formerly belonged to the Malherbe Collection. Alfred Malherbe, judge of the civil tribunal in Metz, so well known as an ornithologist, got it in 1842 through the agency of Herr Reinhardt, Counsellor of State, from the Royal Zoological Museum of Copenhagen.[2]]

49 and 50. *Munich, Germany.*—These are, and have been for upwards of twenty years, in the Zoological Museum of the Royal Bavarian Academy of Sciences. They were first publicly mentioned by Preyer in the "Journal für Ornithologie," 1862, p. 78 and p. 119. One of them, namely, that one which was formerly in the collection of the Duke of Leuchtenberg at Eichstädt, has up to this time been considered as originating from Greenland, and might as such reasonably excite interest, if we take into consideration the very small number of known Greenland specimens. But from information which I have derived from Dr. Pauly of Munich, *both* specimens are clearly of Icelandic origin. It appears that both the specimen formerly belonging to the Duke of Leuchtenberg and the other Munich specimen were originally bought by Michahelles from one and the same individual in Copenhagen (according to Steenstrup, Professor Reinhardt). The negotiations were conducted through Dr. Kuhn, who was then (1833) living in Nuremberg. The one bird cost 200 florins, the other 50. Dr. Pauly gives the following additional information regarding these two birds : "One bird (the Leuchtenberg one) bears the ticket 'Alca Impennis, Iceland, H. v. L.'; the other is marked 'Alca Impennis, L., Polar Sea, 1836.' The one bird is stuffed standing, the other sitting with the legs and tail lying straight up. Both wear the summer plumage, as Naumann has depicted it ('Naturgeschichte der Vögel Europas,' plate 337, fig. 1), only the colour on the back is not black but brown ; and, moreover, in the standing specimen, brown of a pronounced but dull hue, in the sitting one of an indistinct but glistening hue. Both birds are dressed on the belly-sewing system, and are in good preservation. The sitting bird is, in spite of its posture, about four centimetres higher than the standing one. Both are, according to Naumann, old birds, for both have on the upper part of the beak 7 ridges, on the lower 10–11."
[Blasius conjectures that the sitting bird is a cock, the standing one a hen.]

---

[1] Now in the Public Museum, Milan. See page 81.—S. Grieve.
[2] Professor Steenstrup, in a letter dated 16th March 1885, favours us with the following information :—"The specimen sent to Malherbe came from Iceland in 1831."—S. Grieve.

51. *Naes,* near *Arendal, Norway.*—[Prof. Newton, in *Ibis,* 1870, p. 248, calls attention to this fine specimen, which is in the private collection of Herr Nicolai Aal, proprietor of the iron works at Twedestrand. According to Robert Collett it is of Icelandic origin.[1]]

52. *Neuchatel, Switzerland.*—The beautiful full-grown specimen in the Musée d'Histoire Naturelle of this town was first brought, so far as I know, under the public notice by Léon Olphe-Galliard in the *Ibis* of 1862, p. 302. By means of the accurate particulars given by Louis Coulon, the director and custodier of this museum, Victor Fatio ("Bull. Soc. Orn. Suisse," tome ii., partie 1, 1868, p. 74) was able to state for certain that it was bought at Mannheim in 1832, from a dealer in zoological wares, named Heinrich Vogt, for the sum of 200 francs. This Vogt very probably got it from the Royal Zoological Museum of Copenhagen, as is conjectured by Prof. Newton ("Bull. Soc. Ornith. Suisse," tome ii., partie 2, 1870, p. 157), on the ground of the fact learned by him from Prof. Reinhardt, that in 1833 the Museum in question sold a skin to Vogt, a dealer in zoological wares in Mannheim. There is a slight discrepancy in the dates. Either there is a misstatement of the figures in one of the two accounts, or we may perhaps assume that Vogt about that time got several skins from Copenhagen.

53. *Newcastle-on-Tyne, England.*—In the Museum of Natural History is preserved the only known specimen of a young Great Auk.[2] It was first mentioned with the epithet young by G. T. Fox in his Catalogue of the Newcastle Museum. Mistaking the accounts given by Dresser in his "Birds of Europe," vol. viii. p. 566, I formerly conjectured wrongly that this was the same bird as that which Pastor Otto Fabricius, during his five years' stay in the district of Frederikshaab in Greenland, was able to observe and to kill in the month of August. As Fabricius mentions this fact ("Fauna Groenlandica," 1780, p. 82) so soon after his thorough-going proof that the Great Auk cannot breed near Greenland, Steenstrup, in mentioning this matter ("Videnskabelige Meddelelser" for 1855, Copenhagen, 1856–7, p. 33), puts forward with good grounds the conjecture that it was no downy young bird of the Great Auk that Fabricius observed and stuffed, but only a young specimen of some of the other large swimming birds. The principal grounds adopted by Steenstrup for this opinion were (1) that the Great Auk used to breed so early that the young ones, fitted for swimming and diving, betook themselves to the sea as early as the middle of June, and certainly not first in August; (2) that even if an egg were stolen, no second egg was ever laid in the

---

[1] In a letter to me, dated 7th August 1884, Professor Wilh. Preyer, Jena, says—"This place (which lies 6 German, *i.e.* 27 English, miles from Arendal) should be spelt *Nees.* The proprietor's name should be spelt *Aall.*" A complete account of this specimen has been recently given by Robert Collett in his work "Uber Alca impennis in Norwegen" (Mittheilungen des Ornithologischen Vereins in Wien, 1884, No. 5–6). According to him, the specimen was got from the museum of Copenhagen, in exchange for a bear-skin, some time between 1840 and 1850. It bears summer plumage.—W. Blasius.

Professor Steenstrup informs us, 16th March 1885, that "Reinhardt sent the specimen to Aall in 1845."—S. Grieve.

[2] In the Bohemian National Museum at Prague there is a stuffed skin of a Great Auk supposed to belong to a young bird. However, it appears to have belonged to an older bird than that preserved at Newcastle-on-Tyne. See pages 73 and 78.—S. Grieve.

same year; and (3) that the inside of the young bird killed by Fabricius contained exclusively vegetable food, which was strange to the Great Auk.

But let us even put out of sight these doubts of Steenstrup's—doubts that are besides shared by J. T. Reinhardt (*Ibis*, 1861, p. 15); still I hold it impossible to maintain the identity of the presumed specimen of Fabricius with that now in the Newcastle Museum, for the simple reason that the latter represents a later stage of development. Fabricius represents his specimen as a few days old, and covered only with a grey coat of down; whereas the Newcastle Museum Great Auk, according to all the accounts that have been given of it, possesses already an eye-mark spotted black and white, as well as two or three furrows on its bill—both of which signs point to a longer life. Another fact that leads to the conclusion that this Newcastle bird was not so very young, is the fact that Mr. John Hancock was able to take the bones out of the body with a view to their preservation separately. Before that could be done, the process of ossification must have advanced considerably.[1]

Professor A. Newton, who was formerly inclined to trace back this specimen to Fabricius, though not indeed to the identical bird mentioned above as having been killed by Fabricius, wrote to me on the 25th November 1883, acknowledging the possibility of a different origin, and stating that if it can no longer be held to have proceeded from Greenland and Fabricius, then its original home must probably have been Newfoundland, with which also Tunstall, the original possessor of the bird, as indicated by the title of Fox's catalogue, had connections.

    54. *Newcastle-on-Tyne, England.* [This one, along with the corresponding egg, is at present in the private collection of Mr. John Hancock; but Professor Newton informs Blasius that Mr. Hancock is about to transfer, or has transferred, his private collection to the museum of the town, which will thus now be in possession of two skins.[2] This skin and egg were bought by Hancock through the agency of Mr. John Sewell, of Newcastle, from Mechlenburg, of Flensburg. Mechlenburg, in a letter written to Hancock at that time (April 1844), said that that skin and egg, along with another skin and egg, had been got by him one or two years before from an island off Iceland (he said off the north-east coast of Iceland; but that is manifestly erroneous). Now though the investigations made by Preyer in Iceland differ as to their results from those of Wolley and Newton in many particulars, yet they agree with them in this, that the few specimens of the Great Auk obtained at the beginning of the decade 1840–50 did *not* find their way into the hands of Mechlenburg, and that during the second half of the decade 1830–40 no Great Auks were obtained at all. It is, therefore, clear that these two specimens of Mechlenburg (one of them afterwards Hancock's and the other Champley's) must have been obtained at an earlier time; and probably it is to these two birds that Dr. Nils Kjärbölling refers (though he is somewhat wrong as to the date) when he says ("Ornithologia Danica, Danmarks'

---

[1] On the 16th March 1885, Professor Steenstrup writes us:—"The very wild speculations on the young bird of Fabricius, I am very sorry to see promulgated again and again. This young bird of Fabricius has really nothing to do with *Alca impennis*."—S. Grieve.

[2] These remains of *Alca impennis* are now in the Newcastle-on-Tyne Museum.—S. Grieve.

Fugle," 1856, p. 415):—"The apothecary Mechlenburg, of Flensburg, possesses a pair of birds which were killed in 1829 on the Garefowl Rocks, where they courageously defended their eggs."[1] Presumably these two specimens date from the beginning of the decade 1830–40, and consequently cannot have come from the Geirfuglasker (which sank in March 1830), but must rather have come from Eldey, from which originates, according to Newton (*Ibis*, 1861, p. 390), a bird caught in 1834 during the presence in Iceland of the Crown Prince of Denmark.

For more about this specimen see Newton, "Proc. Zool. Soc.," 1863, p. 438; "Ann. and Mag. Nat. Hist." 1864, vol. xiv. p. 138; *Ibis*, 1865, p. 336, and 1870, p. 260.]

55. *New York, America.*—[The skin mentioned by Robert Champley ("Ann. and Mag. Nat. Hist." 1864, vol. xiv. p. 235), as in the possession of Dr. Troughton, was, after his death, bought by D. G. Elliot, through the agency of Cooke, dealer in natural history wares, for the Central Park Museum of Natural History, New York.[2]] (See Frontispiece.)

56. *Norwich, England.*—[Professor Newton writes to Blasius (17th November 1883), that the specimen formerly belonging to Mr. Lombe has, since his death, been presented by his daughter to the Norwich Museum.]

57. *Oldenburg, Germany.*—Cabanis ("Journal für Ornithologie," 1862, p. 78) was the first to mention this specimen, which is preserved, along with an egg, in the Grand-ducal Museum of Natural History. In a conversation which I had with the present director of that Museum—Herr C. F. Wiepken—he told me that it was bought, about 1840 or '41 (the date could be accurately fixed by an examination of the accounts of Grand-ducal Court-treasury), from Salmin, dealer in natural history wares in Hamburg. It is a fine old bird in summer plumage. Perhaps this is one of the three skins which, according to the investigations of Wolley and Newton, were got in Iceland in 1840 or '41, and were presumably sold, all or in part, by the factor, C. Thaae, to Salmin. It would thus belong to the second last find, whilst the two last birds came in spirits to Copenhagen in 1844. The other two skins of that second last find are now probably in Bremen and in Kiel, or Aarau.

Herr Wiepken was so kind as to furnish me with the measurements of this specimen, as far as that could be done in the case of a stuffed bird. These measurements are as follows, and it is to be noted that in the case of the bill, it is measured as far as the horny covering goes :—Total length, up to the bill, 72·0 centimetres ; tail, 7·6 centimetres ; wing, from the bend to the tip, 16·4 centimetres ; cleft of mouth, 9·8 centimetres ; ridges (?), measured straight, about 8·0 centimetres ; measured along the bend, about 9·1 centimetres ; leg, about 5·3 centimetres ; outer toe, with nail, 7·6 centimetres ; middle toe, with nail, 8·0 centimetres ; inner toe, with nail, 6·0 centimetres.

---

[1] Writing us on 16th March 1885, Professor Steenstrup says :—"Dr. Kjärbölling is not exact enough in his different data concerning the *Alca impennis*, even quite uncritical."—S. Grieve.

[2] Professor Newton, writing to me on 15th April 1884, says that D. G. Elliot, according to his own account, bought the specimen without the feet, formerly in Mechlenburg's possession, for the Central Park Museum in New York. Accordingly, there must now be two specimens in that museum, though this point requires still to be cleared up.—W. Blasius.

58. *Osberton, Nottinghamshire, England.*—[This one forms part of the collection of Mr Foljambe. Attention was first directed to it by Professor Newton (*Ibis*, 1870, p. 258). Professor Newton, writing to Blasius (25th November 1883), says that it can be proved that this specimen was bought in Liverpool in the year 1813, and that hence it can be conjectured that it is identical with that one skin which was sent to England, to one of his friends, in 1813, by Vidalin, Bishop of Reykjavik. If this conjecture is correct, this specimen would come from the last slaughter which took place on the skerries off Reykjanes before Faber visited these places in his famous voyage in June 1821.]

59. *Paris, France.*—In the Muséum d'Histoire Naturelle there is to be found, besides an egg and a fine skeleton, a fine specimen mentioned by Preyer (" Journal für Ornithologie," 1862, p. 77). Professor Newton writes to me (25th November 1883) that, considering that formerly a large number of eggs found their way to France from Miquelon and St. Pierre, the French possessions in North America, one might be inclined to select Newfoundland as the original home of the Paris specimen; but there is on the pedestal the express words, " Coasts of Scotland," and consequently—without, of course, attaching faith to that description—one must in the meantime regard the place of its origin as doubtful. According to Preyer (" Journal für Ornithologie," 1862, p. 119) this specimen has on the point of the beak $\frac{8}{9}$ cross-furrows. Can this Paris specimen be identical with that which Brisson (" Ornithologie," vol. vi. p. 85, I. plate 7) has described out of Réaumur's Collection ? Cuvier states, as Professor Newton has pointed out to me, that Réaumur's Collection was afterwards incorporated in the " Cabinet du Roi," and that again the most, if not all, of the articles remaining over from the " Cabinet du Roi," went to the founding or enlargement of the Museum d'Histoire Naturelle in the Jardin des Plantes. Newton, however, judging from its excellent state of preservation, holds such a great antiquity impossible, and believes rather in a later origin from Iceland.[1]

60. *Pisa, Italy.*—[In the Museo Zoologico del Università. Has breeding plumage. See Newton and Selys-Longchamps in *Ibis* for 1870, p. 258 and p. 450.]

61. *Philadelphia, America.*—Prof. Newton, who saw it himself, was the first to call public attention to this specimen, which is in the Philadelphia Academy of Natural Sciences. Its European origin appears certain. For in the great work, " Birds of North America," by Spencer F. Baird, John Cassin and George N. Lawrence, Cassin of Philadelphia, who wrote the section " *Alcidœ*," says :—" We have never seen a specimen of American origin," and adds that the only two known specimens in American museums were that of Audubon (which clearly he had never seen), and the one that had evidently been in his own keeping in the collection of the Philadelphia Academy of Natural Sciences.

There are also, in the same collection, for which they were procured through the agency of Dr. Wilson, the two eggs belonging formerly to Des Murs, or at least one of them.

---

[1] During the spring of 1884 Professor Wh. Blasius heard of another skin existing in Paris ; see p. 79.—S. Grieve.

62. *Poltalloch, Argyleshire, Scotland.*—Prof. A. Newton informed Prof. Blasius lately that a skin existed in the collection of Mr. John Malcolm at the above address, and also an egg. Both the egg and skin were bought from forty to fifty years ago from Leadbeater in London.—S. Grieve. See p. 94 ; also App., Egg 54, p. 32.

63. *Poughkeepsie, New York, America.*—[In Vassar College. For the origin of this specimen see Prof. Newton, in *Ibis*, 1861, p. 336.]

64, 65. *Prague, Austria.*—[In the Bohemian National Museum are preserved two speci- mens, both of which, as the director of the Museum, Prof. Fritsch, writes to Blasius (25th Nov. 1883), are represented in his work, " Vögel Europa's " (Birds of Europe), table 59, figures 8 and 9. The first is entered in 1854 catalogue of the Museum, as " Gift of the King of Denmark to Baron Feldegg.—Polar Sea." In all likelihood it comes from Iceland, and it is probably identical with the bird killed on Eldey in 1834 that came directly into the possession of the Crown Prince of Denmark (afterwards king), who was then staying in Iceland (see A. Newton, in *Ibis*, 1861, p. 390)—though indeed in that article it is stated that that skin came into the possession of Mechlenburg of Flensburg—and indeed the King of Denmark could have had many other opportunities of presenting a specimen to Baron Feldegg, namely, from those that were sent to Professor Reinhardt and the Copen- hagen Museum. This specimen is an old bird, and is in spring plumage.[1]

The second specimen is said by Prof. Fritsch to be a young bird in early autumn plumage, although he admits that this is doubtful. He has represented it in a plate in his " Vögel Europa's " as having a white throat and perfectly dark cheeks, but no white eye- spots, unlike the young bird in winter plumage, whose picture is given by Dresser and Naumann. In 1863 (when Fritsch first called public attention to it), and for some time pre- viously, this bird belonged to the well-known Woboril Collection. Afterwards it passed, with the whole Woboril Collection, into the hands of Anton Richter, a sugar-refiner in Prague, and at a still later date it was acquired for the Bohemian National Museum.]

66. *St. Petersburg.*—[This one is in the Zoological Museum of the Imperial Academy of Sciences. It was first mentioned, though only incidentally, by C. F. Brandt, in the " Bullétin scientifique publié par l'Académie des Sciences de St. Petersburg," tome ii., 1837, p. 345. This C. F. Brandt, during a journey to Germany in 1836, made numerous pur- chases of birds from a dealer of the same name in Hamburg (see the same work, tome i., 1836, p. 176), who just at that time had got several specimens of the *Alca impennis* from Iceland. Perhaps, then, this St. Petersburg specimen was bought on that occasion.[2]]

67. *Scarborough, England.*—[This one belongs to Mr. Robert Champley. Regarding this one Professor Newton wrote to Blasius a short while ago :—" Mr. Champley has only a

---

[1] According to a letter from Professor W. Preyer, of Jena, dated 17th August 1884, it has $\frac{7}{11}$ furrows on the bill.—W. Blasius.

[2] In May 1884 I myself examined the specimen in St. Petersburg. It has the summer plumage, with two large spots on the breast. On the bill it has $\frac{9-10}{11}$ furrows. In confirmation of my conjecture, I found it marked " Brandt…Island."—W. Blasius.

single skin of the *Alca impennis.*   It was bought, along with an egg, direct from Mechlenburg (as I think, in 1860).   I have good reasons for believing that both skin and egg came originally from Iceland."]

68. *Stockholm, Sweden.*—[The Zoological Section of the National Museum of Natural History contains a specimen, to which, so far as I (Blasius) am aware, public attention was first called by Professor Newton, who saw it there himself.   Blasius was formerly in doubt as to whether this specimen had been procured from Iceland, by way of Copenhagen, between 1830 and 1840, or was the one which in 1817 existed in the collection of Gustav von Paykull, Counsellor of the Swedish Chancery; but a few days ago, Professor Newton informed him by letter that it is not the Paykull specimen, regarding the fate of which he could, when in Sweden in 1867, learn nothing, either in Stockholm or in Upsala, and regarding which Sundevall himself was quite uninformed.]

69. *Strassburg, Germany.*—[This noteworthy specimen is in the Town Museum of Natural History, which is housed in the spacious halls of the Academy.   It is first mentioned, so far as Blasius knows, by W. Preyer in the "Journal für Ornithologie," 1862, p. 78. He says of it there:—"A very damaged specimen, with an artificial white (!) upper jaw. It is the worst example I know."   This specimen was almost miraculously saved during the bombardment of 1870 (see *Ibis* for 1870, p. 518).   The present director of the Museum, Dr. Döderlein, sent Blasius at his request the following interesting information about it (1st December 1883):—"The specimen in the Museum of Natural History bears on it no data relative to its origin, mode of acquisition, &c.   Neither can any information be got out of the catalogues, which for the most part are quite incomplete and unreliable.   Perhaps there are, hidden among the archives of the Museum, some facts that may at some future time cast light upon its history.   Our specimen is in a truly pitiable condition.   Its head, wings, and posteriors reveal suspiciously bare places; on the lower jaw the whole horn-sheath of the bill is wanting; the rest of it is tolerable; the upper jaw is genuine; the feet are very well preserved.   In its markings it corresponds completely with the description given of the summer-birds.   The upper bill reveals seven furrows of nearly the same depth. Length (from posteriors to crown), 61 centimetres; length of head, 16 do.; from eye to end of bill, 12 do.; ridge of the bill (measured straight), 8·5 do.; hand portion of the wing (to the tip), 16·5 do.; tarsus, about 5·5 do.; outer toe, 8 do.; middle toe, 8·1 do.; inner toe, 6·5 do.   The height of the bird is 57 centimetres.   We have no eggs nor parts of skeletons."

Before Blasius received the above information from Dr. Döderlein, Professor Newton wrote to him on the 25th November 1883:—"The specimen of Strassburg has probably behind it the history of an existence of like length to that of the Newcastle-on-Tyne specimen.   It was given by P. S. Pallas to Professor Dr. John Hermann, who then worked in Strassburg.   It is, as early as 1776, entered in a catalogue, and that under the denomination 'Northern Seas' (Mers du Nord).   According to the French geographical nomenclature of that time this may signify Newfoundland.   Thus the Strassburg specimen is perhaps to be regarded, along with the Newcastle-on-Tyne specimen, as the only known specimens of American origin."

The preservation in the best possible state of a specimen like this—so interesting in all probability from a historical point of view—is much to be desired. Edmond de Selys-Longchamps mentioned in 1876 that whilst it was in bad condition, that condition could, in his opinion, be improved without diminishing the scientific value of the specimen ; for the essential parts were not imperfect, but only a small portion of the small feathers, black and white, were wanting, and these could easily be supplied from the feathers of an allied species. ("Comptes-rendus des séances de la Société entomologique de Belgique," 1876, 7th Oct., p. lxvii.)]

70. *Stuttgart, Germany.*—This specimen is in the Royal Cabinet of Natural History. Newton, in *Ibis* of 1870, p. 258, mentions it, on the authority of a letter from Dr. Krauss, dated 7th November 1867. E. de Selys-Longchamps mentions in 1876 ("Comptes-rendus des séances de la Société entomologique de Belgique," 1876, 7th Oct., p. lxx.) the fine state of preservation in which this specimen is, and states expressly that it is in breeding plumage. In a kind letter of the 24th November 1883, Dr. Krauss informs me that the Stuttgart specimen came from the collection of Baron John Wilhelm von Müller, Kochersteinsfeld, Württemberg (*ob.* 1864 ; well known as a scientific traveller, and for many years a member of the German Ornithological Society), and that it was purchased in May 1867 along with a number of other stuffed birds. Dr. Krauss goes on to put forward the conjecture, and to adduce reasons for it, that this Great Auk was a gift from the Zoological Museum of Copenhagen, in return for the many valuable objects brought back from the Baron's North African and other travels, and presented by him, amongst other museums, to that of Copenhagen. He goes on to say, " According to Naumann, our bird is in breeding plumage. Its sex was not noted. At any rate, it is an old bird. Except the white spots in front of the eyes (the left spot, however, does not quite reach the eye), the head has no trace of any whitish or brightish markings, and it is like the upper side, black with a brownish tinge. The under side is (with of course the exception of the throat) perfectly white, only somewhat greyish under the wings. The bird is (measured from the back end of the bill along the back to the tip of the tail) 76 centimetres long, some 8 of which are taken up by the tail. The other measurements are—rictus (in straight line), 9·8 centimetres ; culmen (in straight line), 7·7 centimetres ; do. (measured along the bend), 8·4 do. ; the wings full, 16 do. ; tarsus, about 6·5 do. ; outer toe with nail, about 8 do. ; middle do., 8·3 do. ; inner do., 6.3 do. The front side of the upper beak has 8–9 furrows, of the lower beak, 11–12. We have no bones or eggs of the Great Auk." [1]

In a later communication (12th Dec. 1883), Dr. Krauss says : " Our *Alca impennis* is certainly a fine specimen ; but it is badly stuffed. Consequently I have resolved to have it re-stuffed. In the course of opening it up it has become evident that the skin was cut up under the right wing, and was otherwise badly treated. It seems to have been re-stuffed already ; still the skin has kept its condition well."

---

[1] On 16th March 1885 Professor Steenstrup wrote us as follows :—"The conjecture is quite right. As directors of the then Royal Natural History Museum, Professor Forchhammer and I myself presented Baron Müller with this very magnificent specimen, the last of our duplicate skins from Iceland, in order to introduce an exchange of African and Arctic birds ; but we got nothing of the great harvest of African birds collected by the well-known traveller. This was in 1849 or 1850."—S. Grieve.

[Blasius thinks he is justified in confirming Dr. Krauss's conjecture that this bird originally came from Iceland.]

71. *Turin, Italy.*—This one is in the great Museo Zoologico del Università. So far as I am aware, it was first brought under public notice by Robert Champley (" Ann. and Mag. of Natural History," 1864, vol. xiv. p. 235). Edmond de Selys-Longchamps says (*Ibis*, 1870, p. 449) that it is very well preserved, and represents an ordinary breeding plumage.

72. *Veneria Reale, Italy.*—[Regarding this specimen see *Ibis*, 1870, p. 450, 258 ; do. 1862, p. 303 ; do. 1869, p. 393.]

73. *Vitry le François, France.*—[A fine specimen exists here, first brought under public notice by Victor Fatio, who was informed about it by its owner, Count de Riocour (" Bulletin Soc. Ornith. Suisse," tome ii., partie 2, 1870, p. 148).]

74. *Washington, U. S. America.*—[In the Smithsonian Institute. See Newton, in *Ibis* for 1870, p. 259.[1]]

75. *Vienna, Austria.*—This one is in the Imperial-Royal Zoological Court Cabinet. So far as I am aware, it was first publicly mentioned by Wilhelm Pässler in the " Journal für Ornithologie," 1860, p. 60. In 1877 A. von Pelzeln showed, in the " Mittheilungen des Ornithologischen Vereins in Wien " for that year (p. 4), that this specimen came originally from Iceland, and was bought from Frank in 1831. A letter from Professor Newton informs me, in addition, that Frank got this specimen in 1831 from the apothecary Mechlenburg in Flensburg.

76 and 77. *York, England.*—[The Yorkshire Philosophical Society of this city possesses two specimens, regarding which see the reports of the Council of the said Society for 1853, page 9, and for 1866, page 9.]

LIST OF EGGS OF THE *ALCA IMPENNIS.*

1. *Amsterdam, Holland.*—[In the Museum of the Zoological Society, " Natura Artis Magistra." See Robert Champley, in the " Annals and Magazine of Natural History," 1864, vol. xiv. p. 236. Probably of Icelandic origin.[2]]

---

[1] Herr Schlüter, dealer in zoological wares, Halle, told me in September 1884 that Götz got this specimen (which has been cut up under the right wing) from Salmin of Hamburg, and that Salmin got it from Iceland about 1840.—W. Blasius.

[2] According to a later account given by G. A. Frank of London, this egg comes from the quondam Temminck collection in the Rijks-Museum of Leyden.—W. Blasius.

2. *Angers, France.*—[In the Muséum de la Ville. This is one of four eggs which were seen in Brest in 1859, joined on a string. Probably brought by seamen from Newfoundland.]

3, 4, 5. *Angers, France.*—[In the possession of Count de Barracé. Robert Champley mentioned two of them in 1864 (l.c.), whilst Newton was able at a later time to testify to the existence of a third which he had seen there (*Ibis*, 1870, p. 261). These have originally come from Iceland, by way of St. Malo, some time before 1837.]

6, 7. *Bergues-les-Dunkerque, France.*—These two formerly belonged to M. Demeezemaker, and are now since his death most probably in the possession of his son. They were first mentioned and described by Olphe-Galliard (*Ibis*, 1862, p. 302), and afterwards discussed and given in pictorial representations by Ch. F. Dubois (Alphonse Dubois: "Archives Cosmologiques," No. 2, 1867, p. 33–35, plate 3). One of them is larger than the other (it is 126·82 millimetres according to the figure, 125·80 according to Olphe-Galliard), and has on a reddish-yellow ground a great number of broad, irregular dark-brownish black bands, curves and streaks, distributed in pretty equal proportion over the whole surface. The smaller one (117·77 millimetres according to the figure, 115·80 according to Olphe-Galliard) has, on a bright-grey and somewhat greenish ground, only a few small dark spots and (chiefly at the broad end) isolated dark irregular streaks, which unite themselves, for example, in one place into a striking irregular star with five rays (literally, arms).

8. *Breslau, Germany.*—Belongs to Count Rödern. According to R. Champley, who received the information from the deceased Herr Hühnel, the barber, this egg was represented in Thienemann's plates. ("Fortpflanzungsgeschichte der gesammten Vögel," Leipzig, 1845–1856.) It is the upper figure on plate IVC. (*i.e.* 96) of the series, drawn in 1854 or earlier, published in 1856.[1] This egg formerly belonged to the collection of the barber Hühnel, in Leipzig, and was bought from him some time before his death, probably about 1870, by Count Rödern. W. Pässler mentions this egg as to be found in Hühnel's collection, and gives a very short description of it ("Journal für Ornithologie," 1860, p. 59). Hühnel is said to have got 200 thalers (= £30) for it, whereas he had bought it, about 1835, from Fr. Schulz, dealer in zoological wares at Leipzig, for 7 thalers (= £1, 1s.). I am indebted for this information to Herr G. H. Kunz, manufacturer in Leipzig (of the firm C. F. Kunz), who

---

[1] In a letter to me dated 10th February 1885, referring to the above, R. Champley, Esq., says— "Here it is stated that I received information from Hühnel that the Breslau egg was figured in Thienemann's plate. This is not the case. The work contains figures of three eggs—one my own, another Hühnel's, and the third in the Dresden Museum. Hühnel never sent me a drawing from the Breslau egg. The drawing was sent me by Mechlenburg of Flensburg. I copied the drawing, and have it now, and returned him the one he had lent me to copy. The egg is not figured, to my knowledge, and is a very different egg to those figured by Thienemann, besides being a larger one." On the same day the same correspondent again writes :—"Since I wrote you this morning I have looked at the note appended to the Breslau egg, and find I am correct in my statement. 1861 was the year Mechlenburg sent the drawing. He states he got it (*the egg*) direct from Iceland, and sold it to Breslau. From its size and markings, it is not only the largest but the finest marked egg in existence as regards *blotches*, not streaks."—S. Grieve.

wrote me, a short while ago, that about 1835-39 a rich senator died in Hamburg, and left a collection of stuffed German birds, which Schulz, the dealer in zoological wares at Leipzig, bought up, presumably at a very low price, and retailed again. Along with each bird there was an egg. Among them was the *Alca impennis*, with its corresponding egg. This egg fell to Hühnel, who, according to a letter of that time, now in my possession, from Frank, the old dealer in zoological wares in Leipzig, must have been a very enthusiastic collector of eggs. The conjecture that the previous history of this egg can be traced back to Brandt of Hamburg, and from him to Iceland, seems quite justified. According to a report that has reached me, Hühnel is said to have possessed, in the year 1849, no fewer than three eggs of the *Alca impennis*. In that case it would be doubtful if it was the Hamburg egg that is now in Breslau. But according to the evidence of Herr Kunz, who saw Hühnel every day in the office of barber at his father's house, and never knew of more than one egg of the *Alca impennis* in his collection, the report about the three eggs appears to be erroneous. See also pp. 103, 108.

9, 10, 11, 12, 13, 14. *Brighton, England.*—[The deceased George Dawson Rowley possessed at his death six eggs of the *Alca impennis*. As to the first two of these, see Newton in *Ibis* (1870, p. 261). The third belonged formerly to Mr. Labrey,[1] a shipping agent in Manchester, who got it from the deceased Mr. Wilmot. The fourth belonged formerly to Lady Cust. The fifth and sixth belonged formerly to Lord Garvagh.]

15, 16, 17. *Cambridge, England.*—The brothers A. and E. Newton have possessed for nearly a quarter of a century three eggs of the *Alca impennis*, regarding which the first-named gentleman has recently given me the following information. The first came, as did also the second, from Wolley's Collection. Wolley bought it in 1846 from Mr. Beavan. Beavan bought it from Gould, who in turn had bought it from Brandt of Hamburg in 1835. It is evidently this egg regarding which Newton has conjectured, and clearly in the right, that it can be traced back to the booty taken at Eldey in 1834. The second was got by Wolley, in exchange from Wilmot, in 1856. Before that time it had gone through several private collections. It can be traced back to the years 1837 or 1838, when it belonged to a Mr. Augustus Mason. Beyond that time nothing is known about it. The third was bought by A. Newton in 1860 from Mr. Calvert, who said that he got it from the Museum of the United Service Institution, which was then broken up. Farther investigations as to its origin have remained unsuccessful. Some conjecture that this is the egg possessed by Mr. Salmon up to 1860. In that case it would presumably be one of the two eggs which Mr. Proctor got in Durham in 1832, and afterwards sold to Mr. Salmon for £2 a-piece.[2]

18. *Clungunford, Shropshire, England.*—Mrs. Rocke, who presumably is also still the owner of the skin belonging to her late husband, possesses an egg, bought by Mr.

---

[1] In a letter dated from Burslem, 4th April 1859, and addressed to R. Champley, Esq., Mr. Labrey says :—"I have an egg of the Great Auk, which is not by any means a good specimen."—S. Grieve.

[2] This must be a mistake. See Mr. Proctor's letter, page 22 of this work.—S. Grieve.

Rocke in 1869 from Mr. E. Burgh, in whose family it had been for upwards of seventy years. From its age it is probable that it has come originally from Newfoundland.

19. *Croydon, Surrey, England.*—Mr. Crowley got, along with the whole of the egg collection of Mr. Tristram, an egg of the *Alca impennis* (R. Champley in " Ann. and Mag. of Nat. Hist.," 1864, vol. xiv. p. 236). Tristram got it from the late J. de Capel Wise, who, according to Professor Newton, bought it in Copenhagen from Kjärbölling (?). According to one report, two eggs were, as late as 1844, sent from Iceland to Copenhagen. Perhaps Mr. Crowley's egg is one of these two. Its Icelandic origin is very probable.

20. *Dieppe, France.*—The ornithological collection of M. Hardy, who died 31st October 1863, appears, according to all accounts, to be now in the Musée de la Ville. This collection contains, along with a skin, an egg of the *Alca impennis*, which Wolley saw there in 1847 or earlier, Newton in June 1859. Hardy told Newton, on the latter occasion, that he had got the egg along with others from *Newfoundland ;* but as he had before given Wolley his promise that he would try to get him another egg from *Iceland*, Hardy's egg probably came originally from Iceland.

21. *Dresden, Germany.*—[This one is in the Royal Zoological Museum. It formerly was part of the collection of Herr Thienemann, who has given a representation of it in his great work, " Fortpflanzungsgeschichte der gesammten Vögel. One hundred plates of birds' eggs in colours. 1845-56." It is to be found in plate IVC. (*i.e.* 96) in the lower figure to the left. Pässler gives a short description of it in the " Journal für Ornithologie," 1859, p. 59. Its origin is doubtful.]

22. *Düsseldorf, Germany.*—The large private collection, known as the Museum Löbbeckeanum of Herr Th. Löbbecke, up to 1873 apothecary in Duisburg, but now retired from business, contains the collection of eggs inherited by him from his uncle, Friedrich Löbbecke, merchant in Rotterdam, who died on the 29th of February 1856, in the sixty-sixth year of his age, and in it one egg of the *Alca impennis.* Herr Louis Berger, merchant in Witten-am-Ruhr and member for the Landtag, for many years an intimate friend of the old apothecary and zoologist, F. W. J. Bädeker, in the same town, said in a letter to me not long ago, that when he was travelling in Holland along with Bädeker in 1848 he saw this even then somewhat faulty egg at Löbbecke's in Rotterdam, and was told then that Löbbecke had got it from Thienemann. This origin appears to me to be very probable, as the present possessor of the egg writes to me that Friedrick Löbbecke was an intimate friend of Thienemann's, and spent every year a considerable time with him in Dresden. This egg was broken through the awkwardness of a photographer when the present possessor was having it photographed at the urgent request of an English collector ; but it was afterwards so skilfully put together again that the damage can hardly be noticed. The possessor has been able from correspondence to state positively that the egg was bought by Frederick Thienemann from Perrot in Paris at the beginning of 1846. Berger, in a letter to me, conjectures that the picture of this egg has been given by Bädeker, and that conjecture seems to me to be confirmed by the fact that whilst Bädeker's upper figure is taken from

the egg that was formerly Mechlenburg's, but now Robert Champley's, a drawing, description, and note of the size of the Düsseldorf egg given me by Herr Th. Löbbecke answers well to the lower figure. Herr Löbbecke says : " Ground colour of a green dirty yellow tinge ; the roundish grayish-black spots running into one another are mostly at the thick end ; whilst the middle and the thin end have very few spots. Size, 128 millimetres by 75." [1]

23 and 24. *Edinburgh, Scotland.*—[(*Ibis*, 1869, p. 358–360 ; also *Ibis*, 1861, p. 387).]

25. *Hitchin, Hertfordshire, England.*—Belongs to a Mr. Tuke. Is of Icelandic origin. First mentioned by Hewitson in his "Coloured Illustrations of the Eggs of British Birds," 1846. Tuke got the egg from Reid of Doncaster ; Reid got it, on the 23d May 1841, for £2, 6s., from Friedrich Schulz of Dresden, who is presumably the same as Schulz of Leipzig. This information is based on the accounts given by Reid himself to Mr. Robert Champley on the 26th July 1860, as made known to me by Champley in a letter of this year (1884).

26. *Kopenhagen (Copenhágen), Denmark.*—This egg, presumably from Iceland, is in the Royal University Zoological Museum. J. Steenstrup, in a letter to me this year, tells me that no other egg except this one is known to exist in Copenhagen either in museums or in private collections.

27. *Lausanne, Switzerland.*—As Victor Fatio mentions in the " Bull. Soc. Orn. Suisse," tome ii., part 1, 1868, p. 75, two eggs of the *Alca impennis* were found by Dr. Depierre in a drawer about the year 1860. These had probably been acquired by the town of Lausanne at the time when it bought up the collection of Professor Daniel Alexander Chavannes. Chavannes, again, probably got them along with the remains of Levaillant's collection. Fatio has given complete measurements and descriptions of both these eggs ; but as one of them—the smaller and darker—has been disposed of to Frank by way of barter, and by Frank in turn to Lord Lilford, I shall here give only the more important particulars regarding the remaining egg. Its size is 122·5 millimetres by 75·5 millimetres. The two axes intersect at a distance of 44 millimetres from the broad end,—that is to say, at ·359 of the larger axis. The grain of the egg-shell is less developed than in the other egg, hence it possesses rather more lustre. The ground-colour is bright-yellowish, strewn with black and dark-brown spots, which are pretty large, considerably apart from one another, and intermingled with some streaks of the same colours in different directions. The one side is less covered with marks than the other, and at the broad end are more spots, but still without their forming a regular crown. This egg must in colour be especially like the upper figure in Bädeker ; its ground-colour must be like that of the upper figure in Dubois (see Bergues-les-Dunkerque), and its markings like those of the egg in Des Murs' first plate. It probably originates from Newfoundland.

---

[1] In the meanwhile Herr Hugo Klönne, artist in Düsseldorf, has painted this egg for me in oil-colours, and Herr Director Pohlmeyer of Dortmund has prepared plaster casts of it, which show with clearness that the figure by Bädeker mentioned above does not belong to this bird.—W. Blasius.

28. *Leyden, Holland.*—In the Zoological Museum. Newton, writing to me a short time ago, tells me he saw it there in 1860. It probably originates from Frank and Iceland.[1]

29. *Lisbon, Portugal.*—The Museu Nacional (Secçao Zoologico) contains an egg discovered a short while ago among the old contents of the Museum. Ph. L. Sclater called attention to it quite recently (*Ibis*, 1884, p. 122).

30. *Liverpool, England.*—The Museum possesses an egg which, according to R. Champley, belonged to the late Lord Derby. [The thirteenth earl,—*i.e.* the last but one.] This egg was found after his death, in 1851, by Mr. J. T. Moore. Nothing is known as to its origin, but it is one of the most beautiful eggs that exist. Mr. J. Hancock has, as Professor Newton tells me in a recent letter, prepared an excellent coloured plaster cast.

31, 32. *London, England.*—British Museum. These two eggs probably come from Bullock's Collection. At the sale of Bullock's Collection in 1819 two eggs were included in the catalogue (one at p. 31, and the other at p. 131). Both were bought by Leach, then keeper of the Zoological portion of the British Museum, and these are presumably the two eggs now in the Museum. One of them was actually packed in the same box in which Bullock's bird from Papa Westra was; but that does not necessarily lead to the conclusion that the egg came from Papa Westra. To judge from its age, Newfoundland is probably the place of its origin. According to other accounts, these two eggs originally belonged to Sir Hans Sloane, all of whose collections became part of the British Museum.

33, 34, 35. *London, England.*—[In Hunterian Museum. See "Natural History Review," October 1865, and *Ibis*, 1870, p. 261.]
London collection of Mr. Edward Bidwell. See under Weston-super-Mare, App., p. 34.

36. *London, England.*—[Lord Garvagh possessed three eggs, mentioned by Champley in the "Annals and Magazine of Natural History," 1864, vol. xiv. p. 236. Two of them passed not very long ago into the hands of the late Mr. Rowley. It is not known in whose possession the third egg is now.] See also eggs 13, 14, App., p. 26.

37, 38, 39, 40. *London, England.*—[Lord Lilford possesses four eggs; (1) the one which his brother-in-law Crichton got out of the Museum of the Royal College of Surgeons; (2 and 3) the ones found in Edinburgh; (4) one which he bought in the autumn of 1883 from G. A. Frank, dealer in zoological wares, London. Frank at first wanted upwards of £140 for it, but sold it for a somewhat lower price. Frank got the egg, directly or indirectly, by way of barter, from the Museum at Lausanne, as he himself has told me. It

---

[1] This egg belonged to the old Temminck collection. Along with another egg which afterwards found its way to Amsterdam, it was mentioned and described in 1833 by F. A. L. Thienemann ("Systematische Darstellung der Fortpflanzung der Vögel Europás," &c., Part v. pp. 57, 58).—W. Blasius.

is one of the two eggs noticed above under Lausanne. Frank said it was the darker of the two : it must, therefore, according to Fatio's account, be the smaller of the two. The ground-colour is bright yellow-brown, adorned with rather numerous and closely contiguous spots, arabesques, and little streaks of black and blackish tint. These markings follow for the most part the general direction of the principal axis, and running into one another, form a crown close round the broad end. The ground-colour is darker than is reported in the case of any other egg, and the markings put one in mind of those on the egg in Des Murs' second plate. Size, 111 millimetres, by 72·2. The two axes intersect at a distance of 42·5 millimetres from the broad end, or ·382 of the major axis. This egg came, in all probability, originally from Newfoundland.][1]

41. *London, England.*—Belongs to Mr. G. L. Russell, who got it after the death of Mr. Wilmot, its former possessor. Wilmot bought it in 1846 from Leadbeater. A picture of it is to be found in the third edition of Hewitson's "Coloured Illustrations of the Eggs of British Birds," plate 129.

42. *London, England.*—See Newton in *Ibis*, 1861, p. 387. This egg of Mr. Scales' has been lost sight of since 1866. According to a somewhat vague piece of information that has reached me, this egg is now to be found in Dublin. In that case it would be the only Great Auk egg in Ireland. Its original home was probably Newfoundland.[2]

43. *Manonville, Meurthe* [should be *Meurthe-et-Moselle*], *France.* — Baron Louis d'Hamonville purchased, through the agency of M. Dubois, dealer in zoological wares, Paris, the whole collections of Mr. Bond, including a Great Auk egg. Bond got this egg, through the agency of Mr. Gardiner, at the sale of Yarrell's collections. Yarrell got it many years before from a curiosity-dealer in Paris for only a few francs. According to another story, which is traceable to Yarrell himself, he was, forty years before that time, or in other words about the beginning of this century, taking a walk in the neighbourhood of Boulogne, when he met a fish-woman carrying sea-mew eggs. His attention being attracted thereby, he followed her to her house, where he saw hung up on a string four eggs of the *Cygnus musicus*, and in their midst an egg of the *Alca impennis*. He purchased the whole lot at two francs each. This egg is given in the first edition of Hewitson's work on the eggs of British birds, plate 145. Robert Champley has in his possession a hand-drawing of the egg, which he made in June 1860 when the egg was still in the possession of Mr. Bond.

44. *Newcastle-on-Tyne, England.*—Mr. John Hancock bought from the apothecary Mechlenburg of Flensburg, through the agency of Sewell, an egg with the corresponding

---

[1] During the summer of 1884 Lord Lilford purchased a fifth egg, which we refer to at pages 88 and 110.—S. Grieve.

[2] Writing to me on 4th December 1884, Professor Newton says that Mr. Scales died in September 1884, at Brighton, at the advanced age of 90. The son of the deceased had informed Professor Newton that the egg was destroyed by fire some twelve years ago. Fortunately plaster casts of it exist in the collections of Mr. Hancock and Professor Newton.—W. Blasius.

skin in April 1844 or 1845 (that he bought it at an earlier date appears to be a mistake). Whether Mechlenburg got, as he said, the egg and skin shortly before from Iceland may be left an open question. It appears to be certain that in the decennium 1830–9, probably in 1831, they were got on Eldey near Iceland. It is said that Hancock's collections have been given, or are to be given, by way of gift to the Museum of the Philosophical Society. The egg is given in the second edition of Hewitson's work, "Coloured Illustrations of the Eggs of British Birds," on plate 115. It is also briefly described by Pässler in the "Journal für Ornithologie," 1860, p. 59.[1]

45. *Nunappleton, Yorkshire, England.*—[Sir William Milner bought an egg from M. Perrot, a dealer in zoological wares in Paris. He paid 200 francs (= £8) for it. This egg is now in the possession of Sir Frederick Milner.]

46. *Ohinitahi, Canterbury, New Zealand.*—Mr. T. H. Potts formerly owned three Great Auk eggs, which he bought from Mr. Gardiner, senior. In May 1853 he sold two of them at an auction in London to Lord Garvagh, and subsequently took the remaining one to New Zealand. In 1871 he gave an account of it in the "Transactions of the New Zealand Institute" (iii. p. 109.)

47. *Oldenburg, Germany.*—[In the Grand-ducal Museum of Natural History. See regarding it the "Ann. and Mag. Nat. Hist.," 1864, vol. xiv. p. 320. A coloured drawing of it by Sehring, prepared at the Ornithological Congress in Brunswick in 1855, which, however, gives somewhat too small dimensions, is now in the Museum of Hildesheim. There exist also coloured plaster casts of this egg, *e.g.* in the Museums of Brunswick, Hildesheim, &c., as well as in the private collections of A. and E. Newton and Herr Pohlmeyer of Dortmund, who, famed as an egg painter, has himself put the colours on his cast. The measurements of the egg are, according to Herr C. F. Wiepken, 121 millimetres by 75.]

48. *Oxford, England.*—There is now in the University Museum of Natural History the egg that formerly belonged to Sir Walter C. Trevelyan, in whose family it was for over forty years. He got it from Lady Wilson of Charlton House, Blackheath.

49. *Papplewick, Notts, England.*—Mr. Walter has an egg mentioned as early as 1856 by Hewitson in his "Coloured Illustrations." Mr. Walter bought it about 1850 from Dr. Pitman, along with the rest of his collection. Pitman, as Professor Newton tells me in a recent letter, got it from Herr Brandt, of Hamburg. Iceland is its place of origin.

50, 51, 52. *Paris, France.*—In the Museum of Natural History in the Jardin des Plantes there are said to be now three eggs of the *Alca impennis*. One dates from last century, when it belonged to the Abbé Manesse. It probably came originally from Newfoundland. (See Des Murs, "Revue et. Mag. de Zoologie," 1863, p. 4.) The other two were discovered in December 1873 in the Lycée of Versailles. The photographs of these two eggs, in the possession of A. Newton, show that on each of them there is written, "St. Pierre, Miquelon."

---

[1] The collection of Mr. John Hancock is now in the Museum, Newcastle-on-Tyne.—S. Grieve.

Consequently it is certain that they both originate from these formerly French colonies off the coast of Newfoundland. [So Blasius; but St. Pierre and Miquelon are still French colonies.] These are probably the only eggs whose American origin is indisputable.[1]

53. *Philadelphia, U.S.A.*—In 1857 A. Newton saw, in the Museum of the Philadelphia Academy of Natural Sciences, the two Great Auk eggs formerly in the collection of M. O. Des Murs, which were sold to Wilson in 1849. Both eggs were bought from dealers in Paris, the first on the 3d June 1830 from Launoy, for five francs, the second on the 10th May 1833 from M. Bévalt, senior, for three francs. In 1863 Des Murs gave a pictorial representation of these, accompanied by a full discussion regarding them, in the " Revue et Magazin de Zoologie." They have both a ground colour of reddish yellow. The first is covered over the whole surface with broad black dingy brown and bright brown streaks and bands, not very close together, and only at the broad end gathered somewhat together. The second is, on the other hand, covered likewise over the whole surface with narrow streaks and flourishes of bright brown and dingy brown, which at times become thick-like drops, cross one another in many places, form in some places star-shaped figures, and display at the broad end a distinct crown.

Both in the work just quoted, and in his larger work, " Traité Général d'Oologie," Paris, 1860, p. 468, Des Murs says distinctly that he had three eggs of the *Alca impennis*, and had sent them to Philadelphia. The third egg must have been somewhat like the egg of Count Rödern in Breslau and that in the Dresden Museum (these two are somewhat like each other), which Thienemann has given on plate IVC. (*i.e.* 96) of his work (upper figure and left one below), since Des Murs could erroneously conjecture that these two figures represented that third egg of his from two different points of view. It has not yet been clearly proved what has become of this third egg. A. Newton, led by information received by him orally from Cassin, doubts its existence. In a letter to me, Professor Newton says it is reported that one of the two eggs seen by him in Philadelphia has been transferred to the Smithsonian Institution at Washington. This as yet somewhat vague report applies, perhaps, to Des Murs' third egg. In that case the other two will both be still in Philadelphia; but in the meantime I note just one for this place.

54. *Poltalloch, Argyle, Scotland.*—Professor A. Newton has quite recently informed me that Mr. John Malcolm has in his collection a Great Auk egg, as well as a skin not mentioned in the former part of this paper. Both egg and skin were bought, from forty to fifty years ago, from Leadbetter in London. See skin 62, App., p. 21.

55. *Reigate, Surrey, England.*—[Mrs. Wise, who lives in the neighbourhood of this town, has inherited from her father, Mr. Holland, an egg which he bought in 1851 from Williams of London, and which Williams in turn bought from Lefèvre of Paris. Perhaps this is the egg from which Lefèvre caused plaster casts to be prepared, one of which is now in the possession of Professor Victor Fatio, and has been accurately measured and described by him in the " Bulletin Soc. Orn. Suisse," tome ii., part 1, p. 78.]

---

[1] Recently Professor Wh. Blasius has obtained information which leads him to suppose there are now no eggs of *Alca impennis* in the Natural History Museum at Paris. See p. 89.—S. Grieve.

56. *Scarborough, Yorkshire, England.*—The Museum of Natural History here possesses an egg bequeathed to it by the late Mr. Alwin S. Bell. Mr. Bell, in sending the photograph of the egg in 1872 to Professor A. Newton, informed him that he bought it in 1867 from Gardiner, in London. Gardiner only said that the egg came from a collection in Derbyshire, without being willing to give any further information regarding it. Probably Mr. Bell had the egg for some years previous to 1867, since Mr. R. Champley, writing to me a short time ago, informed me that the name given erroneously as Mr. Selwyn, in his former list of 1864, should probably be given as Mr. Alwin Bell.

57, 58, 59, 60, 61, 62, 63, 64, 65. *Scarborough, England.*—These nine eggs belong to Mr. Robert Champley, and were acquired by him in 1864 and a few years preceding that date. Mr. Champley has himself given me the following information regarding them :—

57. Bought from G. H. Kunz of Leipzig. Mr. Champley was himself not well informed regarding the previous history of this egg, since, founding on an evidently misunderstood letter of Pässler's, he traced it back to him. But Herr G. H. Kunz has himself recently informed me that the only egg which he ever had, and which at last he disposed of to Mr. Champley, came from the hands of Herr Th. Schulze of Neuhaldensleben. The son of Herr Th. Schulze, Herr Max Schulze, apothecary and botanist in Jena, has, from papers of his father still in his keeping, been able to inform me that his father sold the egg to Kunz in 1857, for 50 thalers (= £7, 10s.), as I learn elsewhere, and that he got it in 1835 from Herr Fr. Schulze in Leipzig for the price of 7 thalers (= £1, 1s.), along with eggs of the *Podiceps cornutus, Falco Haliaëtus,* &c., and with the accompanying words : " I have had to keep the egg of the *Alca impennis* hidden, as I have been several times asked for it, and it is probably the last I shall get this year."

58. Bought about 1860 from Mechlenburg of Flensburg, along with a skin, the price paid for both being, according to Mr. Champley, £45, and not £120, as given above in accordance with Mechlenburg's memoranda. That Mechlenburg got both skin and egg from Iceland is certain. When he got them, and when they were captured, is still somewhat uncertain. Descriptions of this egg are given by Bädeker in his work on eggs (he also gives a picture of it), and by Pässler in the " Journal für Ornithologie," 1862, p. 59.

59. Bought from Parzudaki of Paris. Parzudaki got it from the Abbé de la Motte of Abbeville. It is described by Bädeker.

60. Bought in Italy from Spallanzani. Accurately described by Bädeker.

61. Bought from Mr. Ward of London, who got it from M. Fairmaire, a dealer in zoological wares in Paris. Size, $4\frac{7}{8}$ inches long, 3 broad. It has a ground colour of dirty white beautifully marked all over with black and brown spots.

62. Also bought from Mr. Ward. Of the same size as 61. The ground colour is dirty white, with dark and brown spots, which form a crown at the broad end.

63. Bought, through the agency of Prof. Flower, from the Royal College of Surgeons. $4\frac{3}{4}$ inches long, $3\frac{1}{4}$ broad. Dark-yellow markings, all at thick end.

64. Bought from the same. $3\frac{3}{4}$ inches long, $2\frac{3}{4}$ broad. Dark yellow ; beautifully marked all over, but somewhat more darkly at the thick end.

65. Bought from the same.   4⅜ inches long, 2⅞ broad.   Ground colour dark yellow, marked all over.

All the nine eggs are in good condition and quite perfect.

66. *Washington, U.S.A.*—See under Philadelphia.

67. *Wavendon Rectory, by Woburn Beds, England.*—Mr. Burney, who lives here, possesses an egg that came from the Royal College of Surgeons, and hence came originally in all probability from Newfoundland.

68. *Weston-super-Mare, Somerset, England.*—The late Mr. Braikenridge, of this place, also bought an egg from the Royal College of Surgeons.   Professor Newton, writing to me, says that it is probably still at Weston in the hands of his heirs.[1]

---

[1] "This egg has recently changed hands, and is now in the hands of Mr. Edward Bidwell, 1 Trig Lane, Upper Thames Street, London, E.C."   He kindly informs the author as follows : "The Great Auk's egg now in my collection is one of the four eggs sold by the College of Surgeons (as per Steven's Sale Catalogue, July 11th, 1865, lot 140).   It was purchased by the late Rev. George A. Braikenridge, of Clevedon, Somerset, from whose sister I recently purchased it."   This egg is now preserved at Mr. Bidwell's residence, Fonnereau House, Twickenham.—S. Grieve.

III.

HAKLUYT'S VOYAGES. London, 1600. Page 1.

The most ancient Discouery of the West Indies, by Madoc, the Sonne of Owen Guyneth, Prince of North Wales, in the yeere 1170 ; taken out of the history of Wales, lately published by M. Dauid Powell, Doctor of Divinity.

AFTER the death of Owen Guyneth, his sonnes fell at debate who should inherit after him ; for the eldest sonne borne in matrimony, Edward or Iorwerth Drwydion, was counted unmeet to governe, because of the maime upon his face ; and Howell, that took upon him all the rule, was a base son, begotten upon an Irish woman. Therefore Dauid gathered all the power he could, and came against Howell, and fighting with him, slew him ; and afterwards enjoyed quietly the whole land of North Wales, until his brother, Iorwerth's sonne, came to age.

Madoc, another of Owen Guyneth his sonnes, left the land in contention betwixt his brethren, and prepared certaine ships with men and munition, and sought aduentures by Seas, sailing West, and leauing the coast of Ireland so farre North, that he came unto a land unknowen, where he saw many strange things.

This land must needs be some part of that Countrey of which the Spanyards affirme themselves to be the first finders since Hanno's time. Whereupon it is manifest that that countrey was by Britaines discouered, long before Columbus led any Spanyards thither.

Of the voyage and returne of this Madoc there be many fables fained, as the common people doe use in distance of place and length of time rather to augment then to diminish ; but sure it is there he was. And after he had returned home, and declared the pleasant and fruitfull countreys that he had seene without inhabitants, and upon the contrary part, for what barren and wild ground his brethren and nephewes did murther one another, he prepared a number of ships, and got with him such men and women as were desirous to liue in quietnesse ; and taking leaue of his friends, tooke his journey thitherward againe. Therefore it is supposed that he and his people inhabited part of those countreys ; for it appeareth by Francis Lopez de Gomara, that in Acuzamil and other places the people honoured the crosse. Whereby it may be gathered that Christians had bene there before the coming of the Spanyards. But because this people were not many, they followed the maners of the land which they came unto, and used the language they foúd there.

This Madoc, arriving in that Westerne countrey, unto which he came in the yere 1170, left most of his people there, and returning back for more of his owne nation, acquaintance, and friends to inhabit that faire and large countrey, wente thither againe with ten sailes, as I find noted by Guytyn Owen.   I am of opinion that the land whereunto he came was some part of the West Indies.

Carmira Meredith filij Rhesi mentionem facientia de Madoco filio Oweni Guynedd, & de sua Navigatione in terras incognitas.   Vixit hic Meredith circiter annum Domini 1477.

(Mr. Hakluyt says that he received the following Verses from Mr. William Camden.)

> Madoc wyf, mwyedic wedd
> Iawn genau, Owen Guynedd;
> Ni fynnum dir, fy enaid oedd
> Na da mawr, oud y moroedd.

The same in English.

> Madoc, I am the sonne of Owen Gwynedd,
> With stature large, and comely grace adorned :
> No lands at home, nor store of wealth me please,
> My minde was whole to search the Ocean Seas.

For further information regarding Owen Guyneth and the name Penguin, see p. 132.

IV.

PROFESSOR J. STEENSTRUP'S REMARKS ON EAST GREENLAND AS AN ANCIENT STATION FOR THE GREAT AUK.

IN a letter dated 30th March 1885 Professor J. Steenstrup has kindly sent us the following information:—"Danells or Graahs Islands may be considered to have been inhabited by the *Geirfugl* in ancient times (300 years ago), if these islands really are the same as Gunnbjornsskjoerne, which perhaps may be the case. Even then it must be remembered that we have only this one visit to the islands recorded (see page 4). During this visit the bird was seen there, and was killed in great numbers. But whether the birds lived there normally, or were accidentally driven to the islands, is quite uncertain.[1] From the expedition sent from Denmark to the eastern coast of Greenland, and to Danells or Graahs Islands, we have as yet not heard anything. It will be of interest to hear if they have met with the remains of the *Geirfugl*.[2]

"In the 'Grönlands Historiske Mindesmærker,' the eastern settlements referred to (*the Osterbygd*) are considered to have been situated on the coast of Davis Strait, and proved to have been there.

"The old settlements of the Norse or the Icelanders in Greenland were all on the southern or most easterly part of the west coast of Greenland, or on the northern or most westerly part of the same coast. Of these settlements, those to the south were named Osterbygd, those on the north Westerbygd. In these localities the remains of numerous ancient settlements have been found.

"The supposition of some authors, supported by 'Mayor' (the voyages of Zeno) and of 'Nordenskiöld,' as to a change of climate, is based on a misunderstanding of old relations (War Bere's)."

When sailing between Iceland and the settlements of Osterbygden, the Gunnbjornsskjoerne are nearly half-way, but near to the east coast of Greenland. We hope that our (Danish) land expedition, travelling along the eastern coast-line, may have reached them during 1884.

---

[1] While attaching the greatest possible value to the opinions expressed by Professor Steenstrup, we would respectfully remark that we think the weight of all the accumulated evidence goes to prove that it was only at its breeding-places that the Garefowl was to be found in such great numbers as appear to have been seen at Gunnbjornsskjoerne.—S. Grieve.

[2] The Danish Land Expedition, 1884.

V.

CORRESPONDENCE REGARDING THE SUPPOSED STUFFED SKIN OF A GREAT AUK OR GAREFOWL (*Alca impennis*, LINN), SAID TO HAVE BEEN SEEN AT REYKJAVIK, ICELAND, BY R. MACKAY SMITH, ESQ., AND PARTY. (See also page 80.)

IN the letter Mr. Smith was kind enough to address to the author on 9th December 1884 (see page 80), he gave the date of his visit to Reykjavik, when he saw this stuffed skin of a Great Auk, as 1858, but finding out afterwards that he had made a mistake he wrote us as follows on 8th April 1885 : "The specimen shot by Mr. Siemsen was seen by several members of our party in the first week of July 1855." On the 13th of the same month, Mr. Smith again writes us :

"Through the Consul-General for Denmark, at the commencement of the Russian War, I received the presentation of four berths on board the war steamship *Thor*. These I presented to the late Sir Henry James, Robert Chambers, John Stuart of Abercromby Place, Edinburgh, and James Laurie, a friend of Chambers'. Sir Henry's chief, Sir John Burgoyne, having to go to the Dardanelles, I took Sir Henry's place. The party of four were joined by the late Alexander Allan, and Robert Allan of Hillside Crescent, Edinburgh.

"The *Thor* arrived in Leith Roads on the 18th of June 1885, when we embarked.

"Of these gentlemen Mr. Laurie is the only one of the five surviving. He writes he saw the specimen of the Great Auk along with a number of other stuffed birds, but evidently he does not remember, for there was no other skin in the house I visited than that of the one in question, which stood in the lobby, the left hand as we entered. He was not aware, nor was I, that this specimen was of much value, which must account for the want of any notes regarding it. Which of the party, all now deceased, accompanied me, I have no recollection, most likely one or both of the Allans and John Stuart.

"I distinctly remember the statement that the bird was shot by Mr. Siemsen a few years previously ; he was the principal storekeeper in Reykjavik, and confirmed this, adding that it was at the skerries (which bear the name of this fowl on the maps of Iceland), and that he believed there was still another specimen there. When I wrote to you last year I had mislaid the memoranda from which I now write (to be exact, the memoranda contain no mention of the Garefowl). 'My impression that the Floors specimen was smaller cannot be insisted upon.'"

On the 18th April 1885, Mr. Smith again writes us to correct a mistake as to the name of the gentleman in whose house at Reykjavik he saw the stuffed skin. He says :—

"Bjarnar Gunnlaugsson was the surveyor for the map of Iceland, occupying him a long series of years. It was at his house I saw the Great Auk, and not at Olsens. Olass Nicolas Olsen directed him as to laying it down on the map. Please correct this mistake arising from my not being able to read Icelandic. I discovered it on questioning an Icelander to-day, who also confirmed my recollections of the locality of the surveyor's house and Gunnlaugsson's being the same. Please to inform Professor Newton, for it is important to know whether Mr. Gunnlaugsson was alive when Professor Newton was in Iceland in 1858. I am writing to Iceland to know when Mr. Gunnlaugsson died, and for other particulars."

We wrote to Professor Newton, and the following is his answer, dated 21st April 1885. He says :—

"I saw Mr. Gunnlaugsson who made the survey of Iceland, and I think more than once.

"Mr. Wolley, I remember, applied to him in regard to the precise position of the Geirfuglasker, but among his papers there is only a short memorandum of what passed between them, and that is not to the present purpose.

"The old gentleman was perfectly aware of our object in visiting Iceland, and it would indeed be very extraordinary if he had had in his house only three years before the skin of a Garefowl, and yet said nothing about it to us ; while, of course, if he had mentioned such a thing, it is impossible for Mr. Wolley not to have noticed it, or for me to have forgotten it."

VI.

REMARKS by R. CHAMPLEY, Esq., ON WHAT SHOULD BE THE ATTITUDE GIVEN TO STUFFED SKINS OF THE GREAT AUK OR GAREFOWL, *Alca impennis*, LINN.

IN a letter to the author, dated 29th April 1885, Mr. Champley says : "Mr. J. Hancock told me there was not one bird correctly stuffed, and he took enormous pains with his own in the Newcastle Museum. All existing stuffed specimens are too stiff in the throat."

Writing us again on 1st May 1885, he says : "Mr. Hancock is the best stuffer in the world ; no one has studied nature closer. All the birds (*i.e.* Great Auks) he has seen have the neck too stiff. His own specimen has the neck *pouched*, so to speak, and not too erect. I think he is correct. He has offered to re-stuff my own bird, but great care is necessary to soften the skin." On the 2nd May Mr. Champley sent us a rough sketch of what he supposes ought to be the correct attitude of the stuffed skins of the Great Auk. There is nothing peculiar about the sketch except the throat, which has in front rather more than half way down from the head a curious projection like a flattened dome. Its gives the appearance to the bird he figures of having a swelling upon its throat, and does not look natural.

VII.

REMARKS BY R. CHAMPLEY, ESQ., ON THE STRUCTURE OF THE SHELL OF THE EGG
OF THE GREAT AUK OR GAREFOWL, *Alca impennis*, LINN.

WRITING to the author on 1st May 1885, Mr. Champley says : "I do not know whether you have ever placed a portion of the shell of the egg under a microscope. If so, you will have noticed the section is not granulated, but transversely laminated, or stratified, if that is the more correct term."

VIII.

IMITATION GREAT AUK EGGS, THE POSSIBILITY OF THEIR BEING PRODUCED IN
PORCELAIN. REMARKS BY R. CHAMPLEY, ESQ.

ON 1st May 1885, Mr. Champley writes as follows to the author :—" I believe attempts will be made to imitate the egg in porcelain similar to the scent-bottles we see in the jewellers' shops. The difficulty will be to prevent cracking during the firing. This may, however, be an advantage as resembling more closely a cracked egg. I believe, however, it will be possible to make a very close imitation, judging from the excellence of the manufacture of the smaller birds eggs as above described.

IX.

CORRESPONDENCE Regarding the Remains of the Great Auk or Garefowl (*Alca impennis*, Linn.), Preserved in the Museum of Natural History at Paris.

THE following information has been received since the greater part of this work has been printed, so there is no reference to this Appendix at pp. 79, 82, 89, and in the Appendix, pp. 20, 31.

As our readers are aware, Mr. R. Champley has been kind enough to look over the proofs of these pages, and observed the statement made by Herr Berger to Professor Wh. Blasius, referred to at p. 89, regarding the eggs of *Alca impennis* said to be preserved in Paris.

Mr. Champley at once wrote to the Director of the Museum of Natural History at Paris, and in a letter to us, dated 14th May 1885, encloses the reply, dated 5th May 1885, of which the following is a translation :

"Sir,—You have been good enough to ask me in your letter of 24th April last what is the number of specimens of the Grand Pingouin (*Alca impennis*), and of the eggs of that species forming part of the collections of the Museum of Natural History. I have the honour to inform you that there is in the closed collections of the Museum one stuffed specimen, one complete skeleton, and three eggs of *Alca impennis*.

"The stuffed specimen came from the coasts of Scotland, and was acquired in 1832. One of the eggs came from the same source. The other two eggs came from the New World, and they were acquired in 1873. They were first found in the collection of the Lyceum at Versailles. (They are shown entered upon the catalogue of the Museum for 1873, Nos. 17 and 18.)

"Thus far from having sold any of the eggs belonging to its collection (which would besides be contrary to the regulations), the Museum, notwithstanding what may have been told you to the contrary, has acquired new specimens.

"The history of the eggs of *Alca impennis* that are to be found in France, has been given by Mons. Daovson Roovby in the Ornithological Miscellanies, &c. &c.

"E. KEMY,
"*Director of the Museum.*

"To Mr. R. Champley,
"*Vice-President of the Philosophical Society,*
"*Scarborough, England.*"

2 A

INDEX.

H

REMARKS

ON

CHART SHOWING THE SUPPOSED DISTRIBUTION OF THE GREAT AUK OR GAREFOWL.

SINCE the Chart was prepared, the Author, from information he has obtained, has seen reason to alter his views with regard to several of the places or localities mentioned in connection with *Alca impennis*, and for that reason he has prepared the following explanatory matter. Owing to the small scale on which it has been necessary to prepare the Chart to take in such a large area of the world's surface, the names of places at certain points are rather crowded. Reference to these remarks, it is hoped, will obviate all difficulties.

Mention is made of several places that have been omitted from the Chart, and there are one or two places that are marked, but which we think ought to be deleted.

(B)

Stands for Breeding-Place.

The following is a list of these. Where a date occurs in addition to the letter (B) it is the last-noted occurrence of the Great Auk at the particular station :—

AMERICAN HABITATS.

CAPE BRETON ISLAND.

 Cape Breton. (See p. 134.)

GULF OF ST. LAWRENCE.

 Bird Islands.—It seems probable that these are the islands referred to at the foot of page 135 as being situated off the Island of Brion and Cape Dolphin, but this does not appear to be quite certain.

MASSACHUSETTS, U.S.

 Cape Cod. (See p. 4.)

NEWFOUNDLAND.

Bird Island off Cape Bonavista.—We have marked this on our map as the island referred to by André Thevet as the island of *Aponars* (see p. 137); but it is uncertain what island is meant, and our marking this island is only conjecture.

Funk Island. (See p. 27) off east coast.
Penguin Islands. (See p. 6) off south coast.

GREENLAND.

Danells or Graahs Islands.—East coast; lat. 65° 20′ N. (See p. 4.)

EUROPEAN HABITATS.

BRITISH ISLES: ORKNEY.
Papa Westra, 1812. (See p. 10.)

OUTER HEBRIDES.
St. Kilda or Hirta, 1821. (See p. 8.)

FARÖE ISLANDS.
Sandoe Island, 1808. (See p. 10.) It is, however, probable the Great Auk at one time bred on other islands of this group.

ICELAND.
Geirfuglaskers.—There were at least three rocks that bore the name of Geirfuglasker, and probably there may have been a fourth. The name itself is confined to Iceland, and may be considered of Icelandic origin.
Geirfuglasker, East of Breiddalsvik, East Iceland (see p. 12).—This rock has unfortunately been named on our chart " Fuglasker," and opposite the name the letter D has been marked instead of B.
Geirfuglasker, Westmanneyar, South Iceland. (See p. 13.)
Geirfuglasker, off *Breidamerkursandr* (see-p. 11), is not marked upon our chart, as it only exists traditionally. It is supposed to have been, or perhaps is still, situated nearly midway between the Westmanneyar and Cape Reykjanes.
Geirfuglasker, off *Cape Reykjanes.*—This skerry disappeared beneath the waves or was destroyed during a volcanic eruption in 1830. It was the principal breeding-place of the Great Auk during the present century. (See p. 18.)
Eldey. (See p. 20.) It was on this skerry that the last Garefowls were killed, at the beginning of June 1844.

(S)

Stands for places where specimens of the Great Auk have been obtained. Where a date is given it is the year in which the last specimen was obtained in that particular locality.

AMERICAN HABITATS.

GREENLAND.

Disco, 1821. (See App. p. 14.) From what Professor Steenstrup says in the note at the foot of the page we quote there is every reason to doubt Disco as a station for *Alca impennis,* L. At the time the chart was prepared we did not know the views of the learned Professor.

Fiskernœs, 1815. (See App. p. 14 under Copenhagen, also foot-note.)

EUROPEAN HABITATS.

BRITISH ISLES.

Farn Islands off Northumberland Coast, last century. (See p. 62.)
Papa Westra, Orkney, 1812. (See p. 10.)
St. Kilda or Hirta, Outer Hebrides, 1821. (See p. 8.)
Waterford, Ireland, 1834. (See pp. 23 and 70.)

GERMANY.

Kiel, 1790. (See p. 23.)

ICELAND.

Eldey. (See p. 20.) The last Great Auks were killed here in 1844.
Geirfuglasker off Reykjanes. (See p. 18.) This skerry disappeared beneath the waves in 1830, not 1829, as was at one time supposed.
Hellerskipna, on Mainland of Iceland, between Skagen and Keblavik, 1821. (See pp. 21, 22.) This place has been omitted from our map by a mistake.
Selvogr, Mainland of Iceland, 1803 or 1805. (See p. 21.) This place has also been omitted from our map by a mistake.
Latrabjarg, 1814. (See p. 21.)

NORWAY.

Frederiksstadt, 1838. (See p. 24, also foot-note.)

SWEDEN.

Marstrand, last century. (See p. 23, also foot-note. p. 24.)
Tistlarna, last century. (See p. 23, also foot-note, p. 24.)

(P)

*Stands for places where the Great Auk may possibly have bred, though the
supposition rests principally on conjecture.*

AMERICAN HABITATS.

NEWFOUNDLAND.

Miquelon, on South Coast.   (See App. p. 31.)
Penguin Rocks, west of Cape Freels, on East Coast.
St. Pierre, on South Coast.   (See App. p. 31.)
Virgin Rocks, east of Cape Race.   (See p. 7.)

EUROPEAN HABITATS.

BRITISH ISLES.

Shetland.   (See pp. 4 and 10.)

ICELAND.

Geirfugladrangr, off Cape Reykjanes.   (See p. 17.)

(D)

*Stands for doubtful localities, where the Great Auk is said to have been seen,
or specimens obtained, or where it is said to have bred.*

AMERICAN HABITATS.

GREENLAND.

Frederikshaab.—This station is marked on account of the young bird said to have
been got here by Fabricius, but as there is little doubt that this young bird was
not an *Alca impennis,* this station should be deleted.   (See p. 72, also foot-note,
and App. p. 1.)

LABRADOR.

Island off the Coast.   We have marked on our map an island in Hamilton Inlet, but
we think this station should be deleted altogether.

NEWFOUNDLAND.

Tail of Newfoundland Banks, 1852.   (See p. 7.)

Trinity Bay, 1853.   (See p. 7.)

EUROPEAN HABITATS.

BRITISH ISLES.

Belfast Lough, 1845. (See p. 23.)
Castle Freke, long strand, west of County Cork. (See p. 23.)
Gourock, Firth of Clyde, near Greenock. (See p. 9.)
Lundy Island, Bristol Channel. (See p. 23.)
Skye, Island of, Inverness-shire. (See p. 24.)

FRANCE.

Brest, beginning of this century. (See p. 23.) Near Dieppe is another locality mentioned. (See App. p. 9.) However, there is little doubt both localities should be deleted from the map as far as the capture of specimens of the Great Auk is concerned.

ICELAND.

Grimsey Island, north of Iceland (not the island of same name in Huna Floi). (See p. 22.)

NORWAY.

Sondmore, near Aalesund, rocks off coast. (See p. 36.) The occurrence of *Alca impennis* here is very doubtful. (See p. 122.)

(O)

Submerged Breeding Place.

ICELAND.

Geirfuglasker off Cape Reykjanes.—This skerry disappeared during 1830. (See p. 18.)

(R)

Places where Remains of the Great Auk have been found.

AMERICAN HABITATS.

NEWFOUNDLAND.

Funk Island, 36 miles north-east by east from Cape Freels. (See p. 27.)

UNITED STATES :

STATE OF MAINE.

Crouches Cave near Portland. (See p. 76.) Not marked on map.
Mount Desert. (See p. 76.)

STATE OF MASSACHUSETTS.

Shell Heap near Ipswich.—"Orton" says, "in shellheaps at Marblehead, Eagle-hill in Ipswich, and Plumb Island." (See note, p. 30.) As we have been unable to discover the exact position of each of these places, we have only marked Ipswich on our map.

EUROPEAN HABITATS.

BRITISH ISLES.

Keiss, Caithness-shire. (See p. 43.)
Oronsay, Argyleshire. (See p. 47.)
Whitburn Lizards, County Durham. (See p. 62.)

DENMARK.

Fannerup Randers, Jutland. (See p. 39.)
Gudumlund, south side of eastern part of the Limfjord, Jutland. (See p. 39.)
Havelse, in Seeland, situated at the southern part of the Issefjord. (See p. 37.)
Meilgaard Randers, Jutland. (See p. 31.)
Solager, in Seeland, northern part of Issefjord. (See p. 39.)

ICELAND.

Baejasker, near Cape Reykjanes. (See p. 41.) From want of room on the small scale this place is only marked on the enlargement of S.W. corner of Iceland.
Kyrkjuvogr, near Cape Reykjanes. (See p. 41.)

PRINTED BY BALLANTYNE, HANSON AND CO.
EDINBURGH AND LONDON.

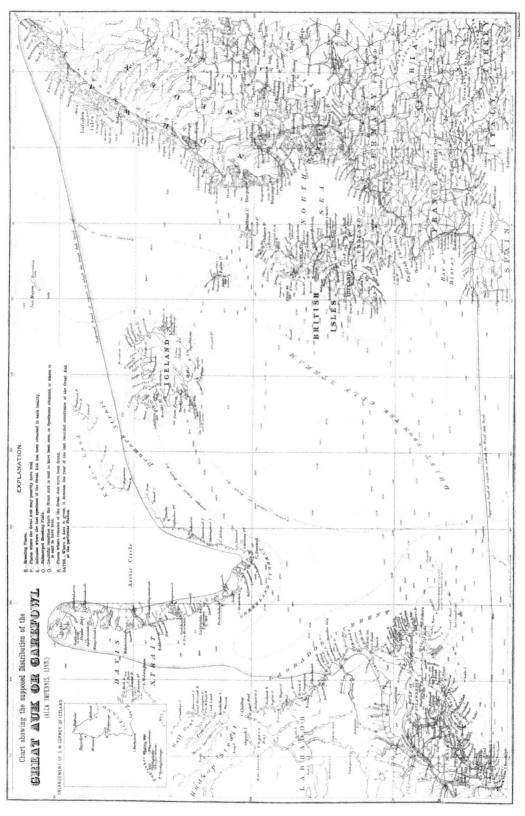

The material originally positioned here is too large for reproduction in this reissue. A PDF can be downloaded from the web address given on page iv of this book, by clicking on 'Resources Available'.